SOLAR ENERGY SIMPLIFIED

FIFTH EDITION

FRANK L. BOUQUET

Systems Co., Inc.
CARLSBORG, WASHINGTON

ISBN 1-56216-122-9 Hardcover Originally $70
ISBN 1-56216-123-7 Paper Originally $40

Library of Congress Cataloging-in-Publication Data

Bouquet, Frank L.
Solar energy simplified / by Frank L. Bouquet. -- 5th ed.
p. cm.
Includes bibliographical references and index.
ISBN 1-56216-122-9 : $70.00. -- ISBN 1-56216-123-7 (pbk.) : $40.00
1. Solar energy. I. Title.
TJ810.B66 1989
621.47--dc20
94-20901
CIP

 FIFTH EDITION

Published By: Systems Co., Inc.
P.O. Box 339
Carlsborg, WA 98324
U. S. A.

Published in the United States of America.

STATEMENT OF CONFIDENTIALITY

PRINTED ON RECYCLED PAPER

PREFACE TO THE FIFTH EDITION

New developments have occurred in solar energy since the writing of the fourth edition. Therefore, there is a need to update the book. Specifically,

- New building designs have emerged.
- There have been new applications to the use of solar energy in agriculture and home uses.
- Information on the biological effects of the sun's rays have been summarized.
- New references have been uncovered and included.

In the field of solar energy, much of the technology is concerned with extensive data, mathematics, and complex equations. Some of the literature requires an extensive engineering background. In this book, as in previous editions, every attempt has been made to reduce the solar fundamentals to simplified form for easy understanding and comprehension.

Because of these developments, the fourth edition has been revised.

Frank L. Bouquet
Carlsborg, Washington
March 1, 1994

PREFACE

The purpose in writing this book is to fill a gap in the open literature in solar energy engineering. Many fine publications exist but none is an introduction to this complex field that is simplified enough for the beginner. An overview that includes important economic factors is needed. The dividual trying to assess personal use of solar energy is confronted with a bewildering variety of options. System analyses of the various technical and economic factors are required in order for one to arrive at a realistic assessment of the benefits of solar energy for a particular case. The result is dependent upon the type of materials used, installation and maintenance costs, the insolation at the geographical site and other complex factors.

The author's goal is to present a picture of the current status of the typical options for solar energy as an aid in making these decisions.

The tabulated solar energy data, as given in detailed source books, such as SOLAR ENERGY HANDBOOK, are not repeated herein but the reader is referred to those references for further information. It is hoped that this approach will be helpful and instructive to both individuals who are learning the field and need an overview as well as "DO-IT-YOURSELFERS".

CONTENTS

Chapter 1

INTRODUCTION

During the 1980's, many applications of solar energy for home and business use have emerged. It is the purpose of this book to summarize these advances by subject for both the layman and engineer alike. Every attempt has been made to reduce the complex technology to simplified form.

Why this book? There are many books available on solar energy. However, they usually treat a specific aspect or require extensive education or experience to understand.

Solar energy technology encompasses many disciplines. It is clear that this book is needed to present an overview for the layman who is trying to decide:

- o Are there any systems I can use?
- o What are their costs?
- o What are their advantages and limitations?

This book attempts to answer these questions for both the Do-It-Yourselfer or the persons planning to subcontract the building of a solar energy system.

- o CHEMISTRY
- o ENGINEERING
- o ELECTRONICS
- o MATERIALS SCIENCE
- o ADHESIVE SCIENCE
- o TAX CREDITS
- o MATHEMATICS
- o ECONOMICS
- o PHYSICS
- o ASTRONOMY
- o LEGAL ASPECTS
- o ACCOUNTING

SOLAR ENERGY

The sun provides the earth with over 50×10^{12} * tons of coal equivalent energy per year and this is greater than 5000 times the present rate of energy consumption.

With the world becoming increasingly dependent upon fossil fuels and the rate of increase of world energy consumption being near 5 percent, the use of renewable energy sources, especially solar appears very attractive.

* 10^{12} is 1 followed by 12 zeros before the decimal point. A very large number indeed.

S Y S T E M   D E S I G N

Experimental work is required to:

Determine the system performance

Determine the system optimization point.

Supplemental information is needed on environmental degradation of performance. Also , data is needed on life-cycle servicing requirements.

The above information is needed to make a full economic evaluation.

METHODS OF CONCENTRATING SOLAR ENERGY

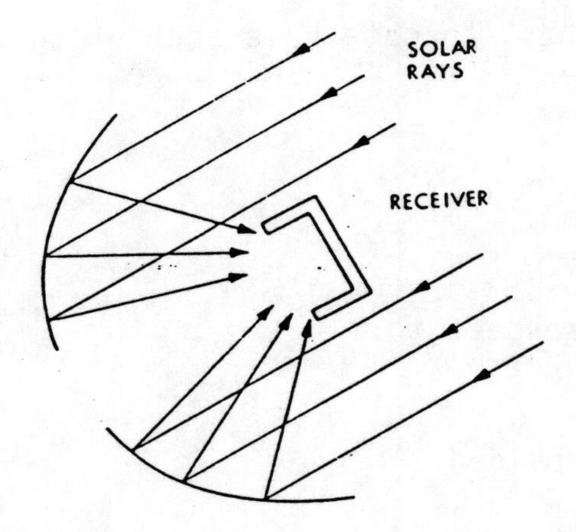

PARABOLIC SHAPED CONCENTRATOR

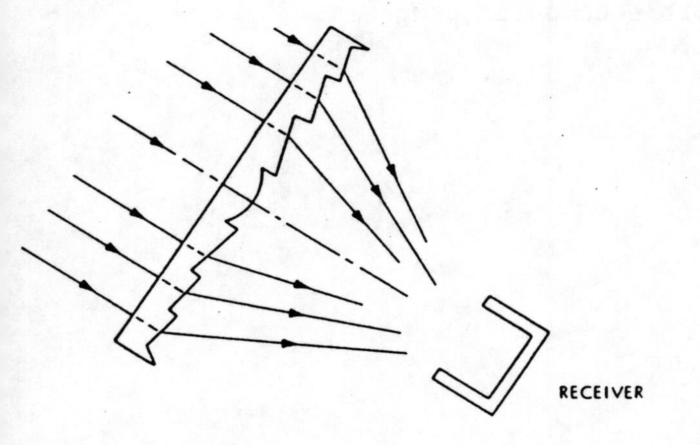

FRESNEL LENS CONFIGURATION

The type of solar heating experiments, to date have been of two types:

WATER HEATING

SPACE HEATING

The parameters are complex with respect to:

Types of design

Proportion of heating loads provided

Typical present and future costs.

These questions are amenable to manual calculations but are best optimized by the use of available computer programs.

A V A I L A B L E D A T A

Seasonal and annual performance data are only available on a limited number of several thousand solar space heating systems now in operation.

Annual costs of the solar heating equipment and installation usually exceed the current values of the energy savings.

The installed systems work well, in general. Air type systems are preferred because they do not boil or freeze.

In summary, there is a tremendous amount of solar energy that impinges on the earth's surface each day. The challenge is to successfully utilize this energy for useful effects for mankind.

Much more remains to be done to exploit this vast resource. Many techniques are possible, ranging from the simple to very complex.

In the following sections, examples of various systems are given, also ranging from the simple to the complex. All systems, in order to be successful, must withstand the effects of sunlight and other environments that may be present.

Some aspects of the environments are treated in the following section.

Chapter 2

INSOLATION

The sun's energy , arriving at the earth's surface, is referred to as "Insolation." The rays arrive as a very diffuse amount of energy.

Flat plate solar collectors receive heat at an average rate of 25-90 watts per square meter.

The successful use of this energy presents a challenge to our ingenuity.

REGIONAL DIFFERENCES IN SOLAR ENERGY AVAILABILITY

Solar radiation availability leads to :

Why some systems are poor performers in some geographical locations, and not in others.

Percent extraterrestrial radiation plots(ETR) show how.

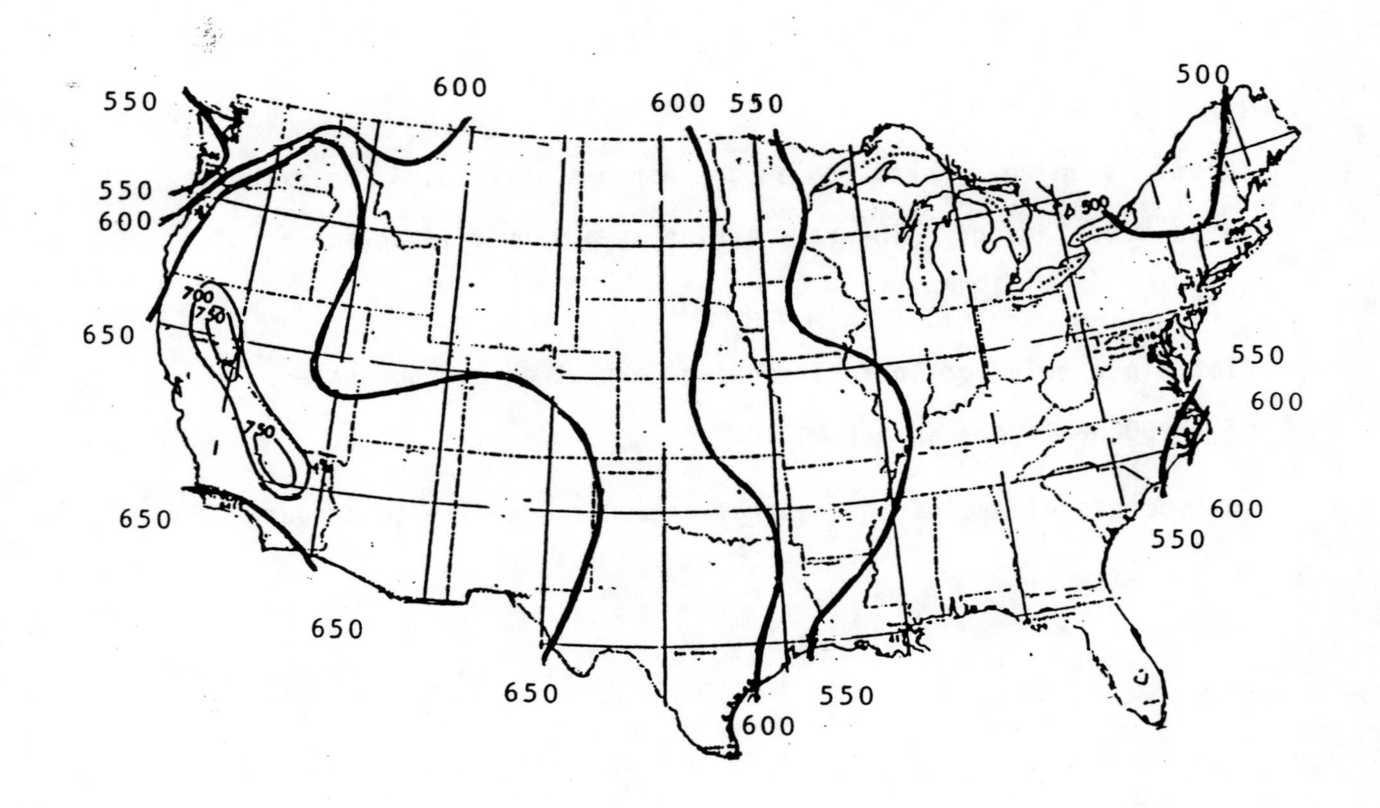

MEAN JULY VALUES ARE SHOWN HERE

VALUES ARE IN LANGLEYS (cal/cm^2)

REGIONAL DIFFERENCES IN SOLAR ENERGY AVAILABILITY

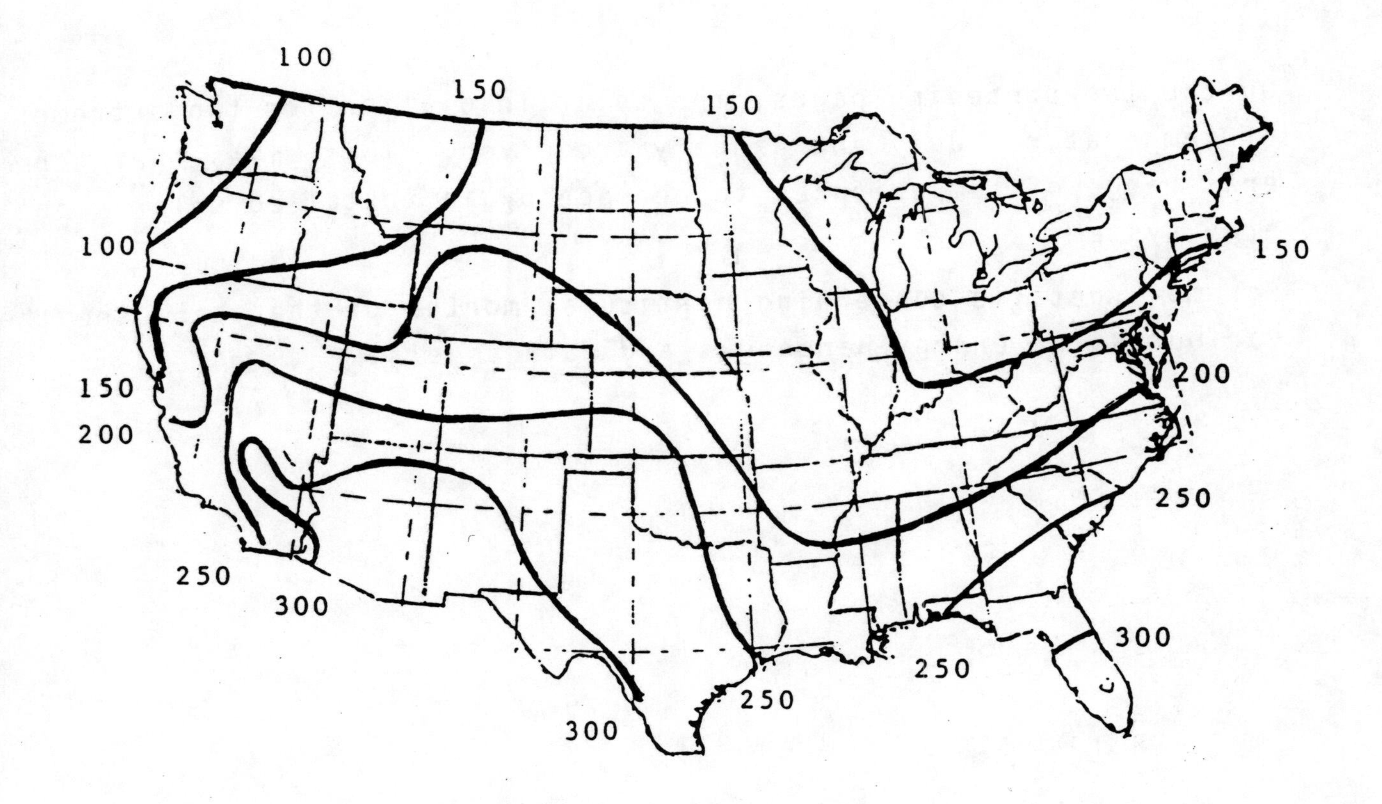

MEAN JANUARY
VALUES ARE SHOWN HERE

VALUES ARE
IN LANGLEYS
(cal/cm^2)

On the two preceding pages, the solar insolation for two extremes of temperature, July and January are given. This gives the general range of temperatures to be seen by an installed solar energy system.

For more details concerning particular months of the year, see the various solar energy handbooks and data sheets.

SOLAR INTENSITY

Although the sun is 93 million miles away, it is huge, some one million miles in diameter, compared with 8,000 miles in diameter for the earth. The intensity at the top of the atmosphere is approximately 1300 watts per square meter decreasing through the atmosphere to approximately 1000 watts per square meter at noon in the summer on earth. Measurements have been made to show the intensity is relatively constant.

The interaction with the atmosphere of the sun's rays results in scattering, absorption and reflection. The amount of energy transmitted depends upon the atmospheric clarity, path length and transmittance.

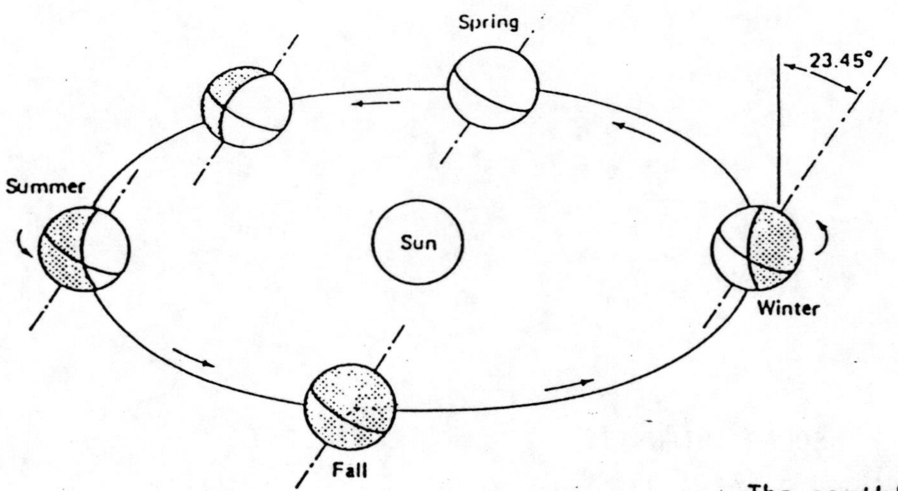

The earth's path around the sun.

Length of the day as a function of north latitude.

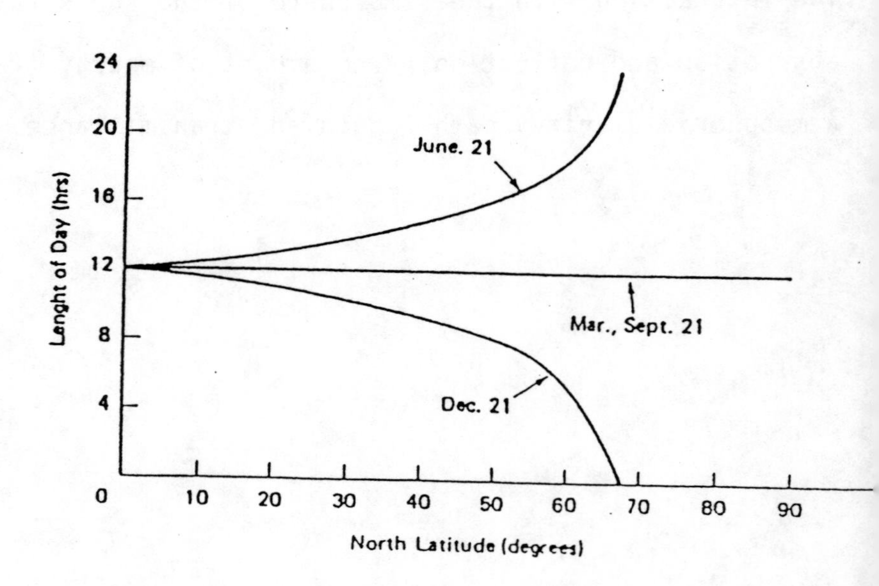

SOLAR SPECTRUM

This figure shows the spectral irradiance at sea level compared to outside the atmosphere with the sun directly overhead.

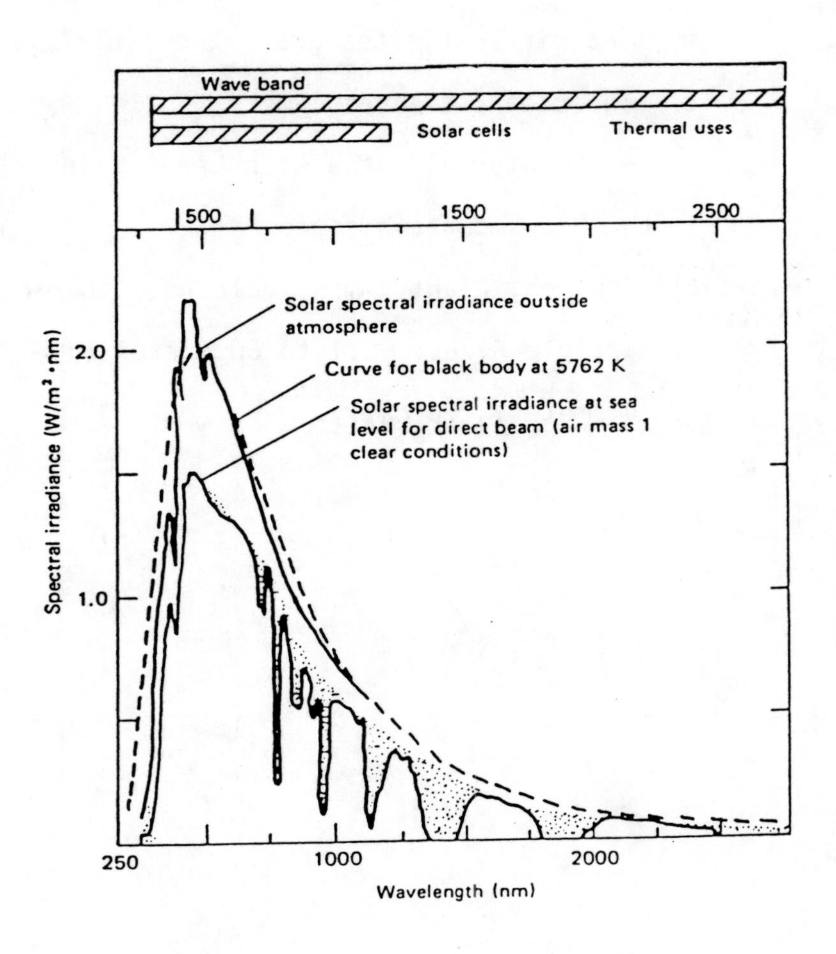

SOILING OF SURFACES

Dust piles up on all exterior exposed surfaces

Research has found that-

- O The dust effect attenuates the solar effect, be it PV or collectors
- o The dirtiest sites are : Industrial locations
- o The cleanest sites
  - Equatorial sites with heavy rainfall
  - High mountain sites

Particle Sources: Plants, man, volcanoes to name a few.

Typical Particle Sizes: 0.1 to 50 microns(i.e. 10^{-4} cm) in diameter, typically.

E N V I R O N M E N T S

CLEANING OF SOILED SURFACES

- FREQUENCY: Depends upon location and the level of efficiency desired by the system.
- HOW ?
  1. Use of commercial cleaners by hand.
  2. Use of high pressure water followed by de-ionized water rinse.
  3. High pressure water with detergent followed by de-ionized water rinse.
  4. Use of tap water with or without prior manual cleaning.
- HOW HIGH PRESSURE ?
  1. Tap water pressure = 80-100 lbs/in^2 may be sufficient for small systems.
  2. Large systems, such as the power tower , may require 500- 800 lbs/in^2.

INSTRUMENTS FOR MEASURING SOLAR ENERGY

Essentially, two types of scientific instruments are used to measure the solar flux:

PYRANOMETER

This type measures hemispherical solar flux.

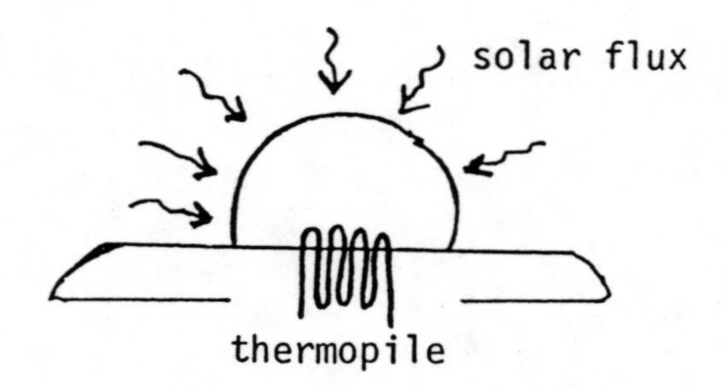

PYROHELIOMETER

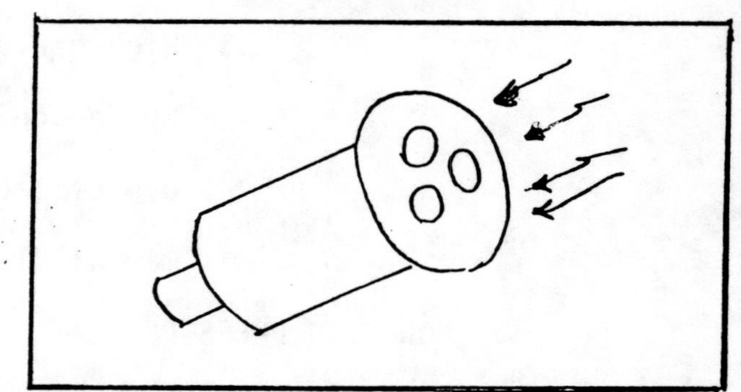

This type measures solar flux at norma
incidence.

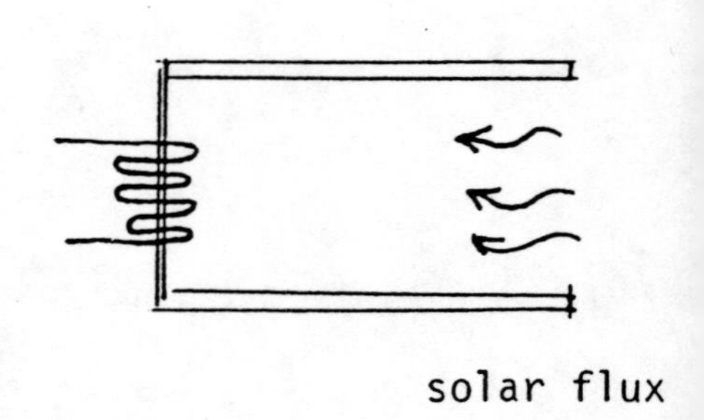

SENSORS TO FOLLOW THE SUN

Sōlar Site Selection Data.

- Lewis and Asso.
  105 Lockwood Drive, SA6
  Green Valley, CA 95945
  916-272-2077

- Sundance Solar
  24 Dickens Circle
  Salinas, CA 93901
  408-422-2000

Sensors

Robbins Engineering, Inc.
751 So. Richmond Road
Ridgecrest, CA 93555
619-375-6882

Chapter 3

PHOTOVOLTAICS

The use of silicon to convert the Sun's energy directly to electricity is a popular application. This technology is a relatively recent achievement, with the first working all being designed in the 1950's.

A major effort was undertaken by industrial companies and the U.S. government to reduce the cost of these devices. Research associated with development of solar cells for satellites and spacecraft was instrumental in advancing this technology rapidly.

An overview is given in this section. For more details on the various methods to design, construct and use photovoltaics, see the References.

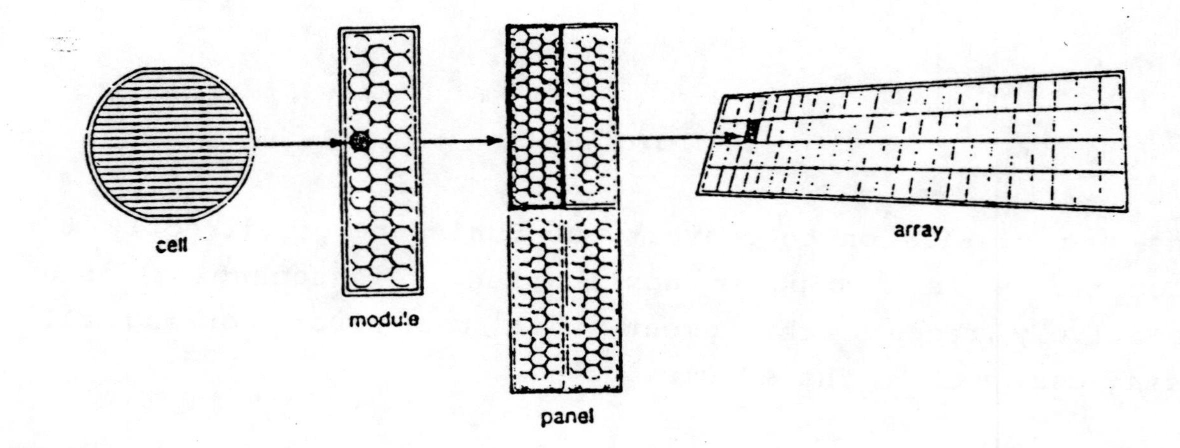
cell
module
panel
array

TYPICAL USES OF PHOTOVOLTAICS

Solar cells are used for power sources for:

- House Power-
  Energy source for home appliances.
- Pumps-
  Energy source for pumping water from the ground for farming or drinking.
- Telephones-
  PV arrays located on top of telephone poles can supply communication energy.
- Remote Power-
  In isolated communities, such as desert towns, PV arrays can be used for lights, radios and other applications.
- Transportation-
  More recently, experimental automobiles and aircraft have been built using PV arrays for energy. As technology progresses, significant improvements are envisioned in the transportation area.

TYPICAL PHOTOVOLTAIC INSTALLATION

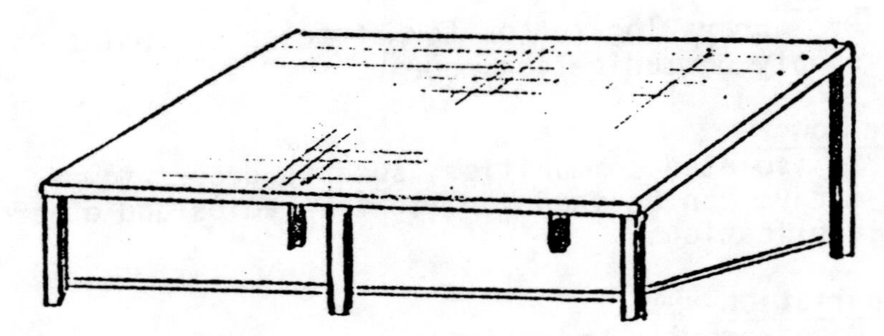

- O Cells must face South in the United States.
- O Although each cell may be only 0.5 volts(typically), the series and parallel arrangement may produce 6-8 volts for every 18 cells.
- O Electrical codes must be used.
- O Long-lived pottants/adhesives such as ethylene vinylacetate(EVA) or polyvinyl butyral(PVB) may be used inside durable glass or plastic outer surface.

SHADOW FACTORS

Shadow factors vary with the geographical position of the collector or PV array.

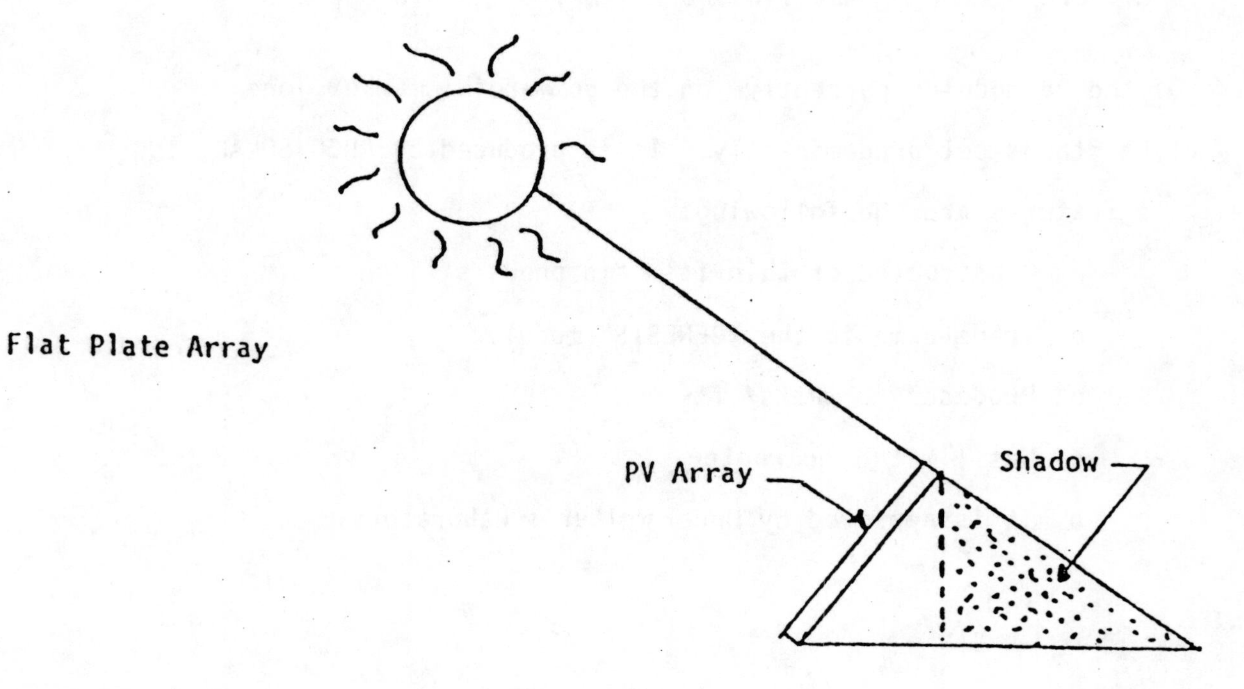

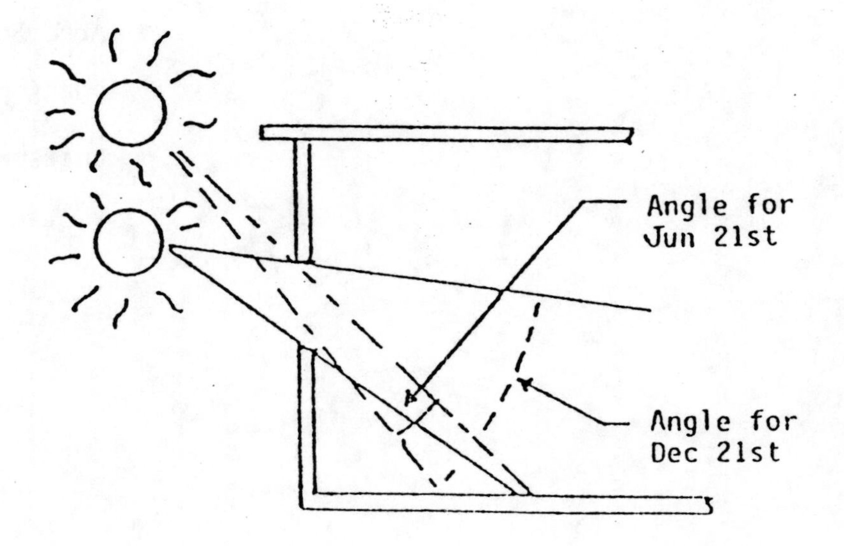

PRICES OF PHOTOVOLTAIC MODULES

Of the PV modules currently on the commercial market, one type stands out predominantly. It is produced by ARCO SOLAR, INC.*
Its features are the following:

- Constructed of thin film amorphous silicon
- Trade name is the "GENESIS" module
- Produces 5 watts/ ft^2
- Has limited guarantee
- It is approved by Underwriter's Laboratory.

* Arco Solar Inc.
20455 Plummer Street
Chatsworth, California 91311

PV PRODUCT AVAILABILITY

Many US companies now produce photovoltaic modules commercially. A listing is given in the following reference.

PRODUCT DIRECTORY

PHOTOVOLTAICS

SOLAR ENGINEERING & CONTRACTING

Vol. 2, No.1, pp. 26-34, January 1983.

PV ELECTRICAL INSTALLATION CODE

Photovoltaic modules, as well as other solar electrical sources, must adhere to the U.S. National Electrical Code. The purpose and objective is as follows:

- Purpose- The purpose of the electrical code is the practical safeguarding of people and property.
- Safety- All parts of the solar PV system must operate at 50 volts or less for personal safety. Components at higher voltages must be shielded against accidental contact.

Energy Flow in a Typical Silicon Solar Cell

The cell can use only 44% of the energy.

16 % of the energy is lost by various processes within the cell.

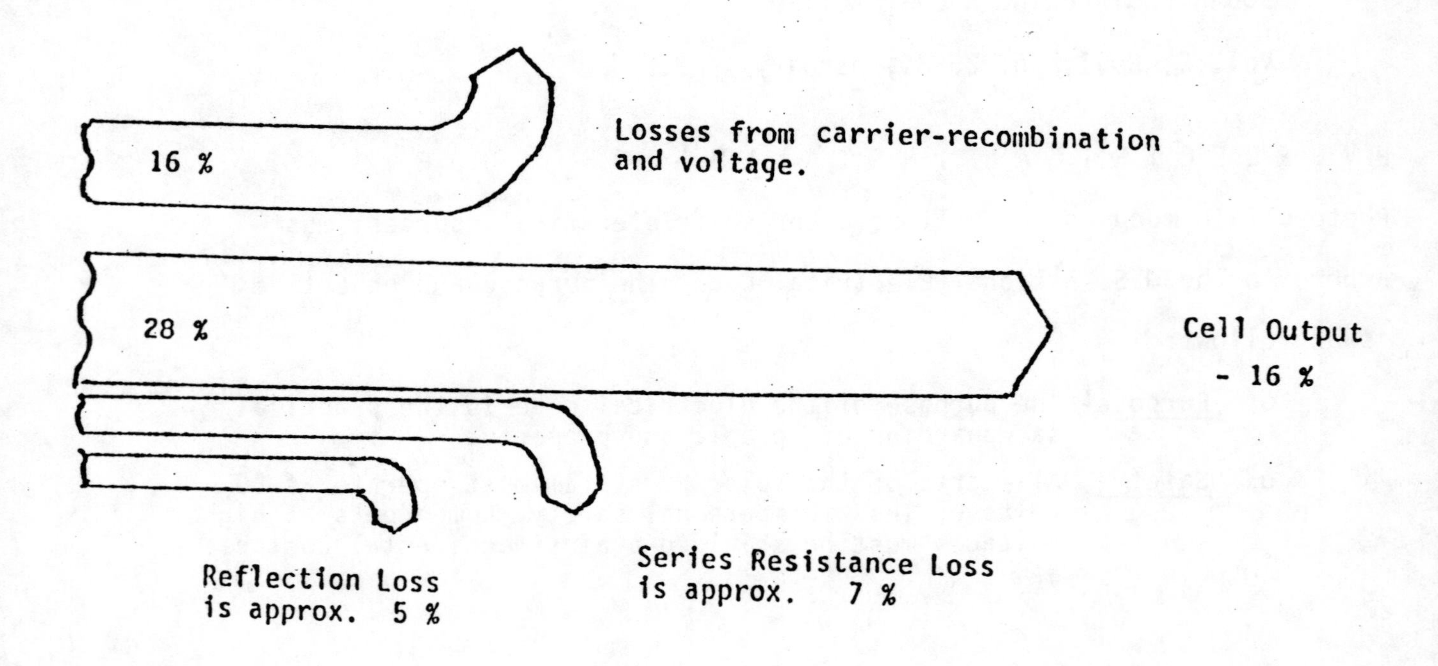

SUMMARY

PHOTOVOLTAIC (PV) ARRAYS

Many types of PV cells exist.
Large 3 inch diameter cells appear to be among the best.

Typical uses are:

- Residential supplemental power
- Isolated power
- Water pumps

Excess power may be fed back into the utility grid to reduce costs.

Current costs are in the 5-10 dollars per peak watt range.

PHOTOVOLTAIC THEORY

- Effect of Photons on Silicon Atoms

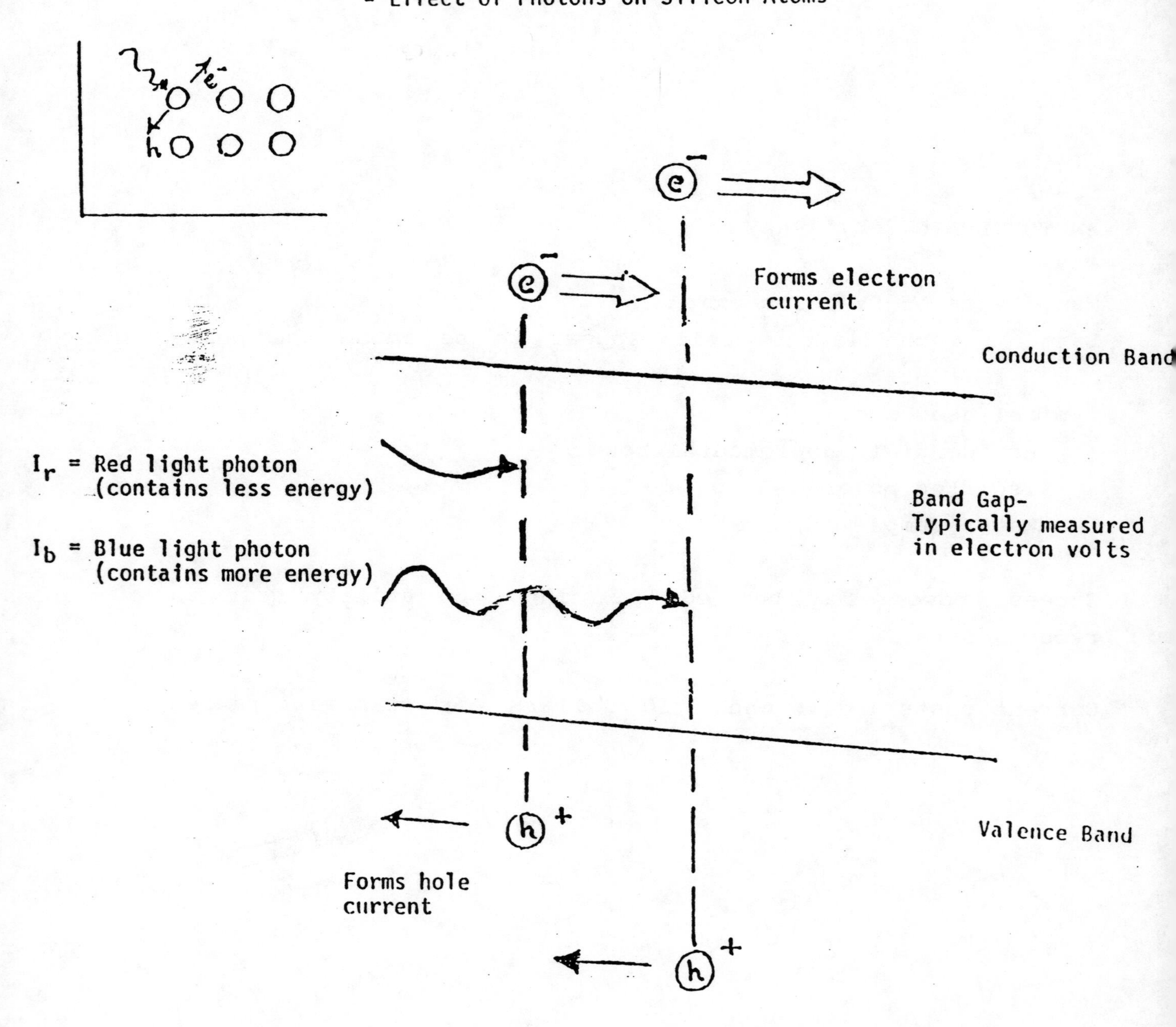

Note: The interactions of the photon with the silicon atoms are shown schematically by the head of the photon arrows.

Holes move to the left when electrons are projected into the conduction band.

PHOTOVOLTAICS

How do they work? There are three main types.

a. SILICON(Si)

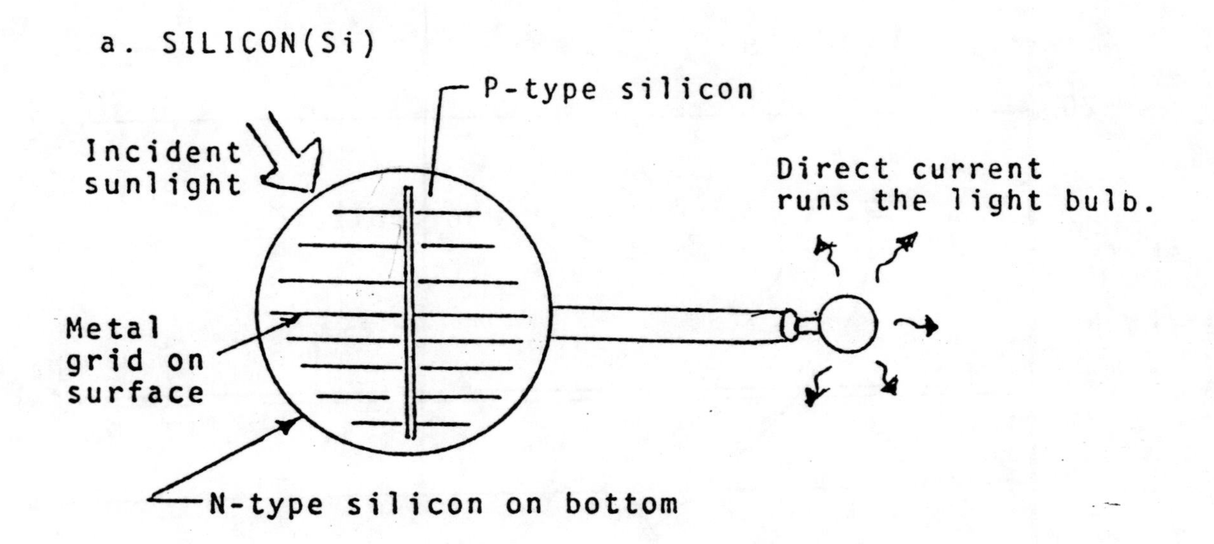

b. CADMIUM SULPHIDE(CdS)

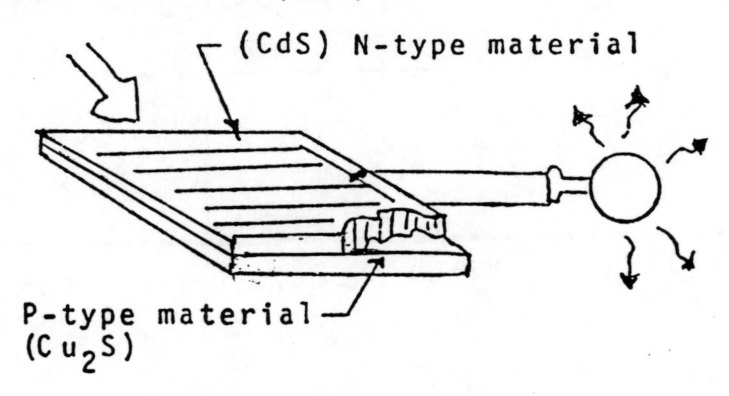

c. GALLIUM ARSENIDE(GaAs)

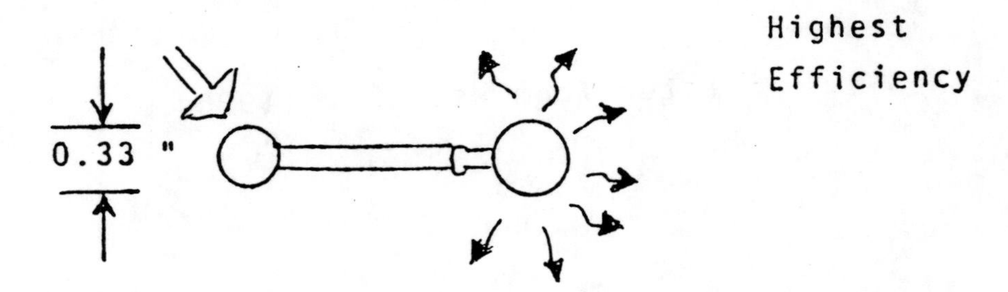

Highest Efficiency

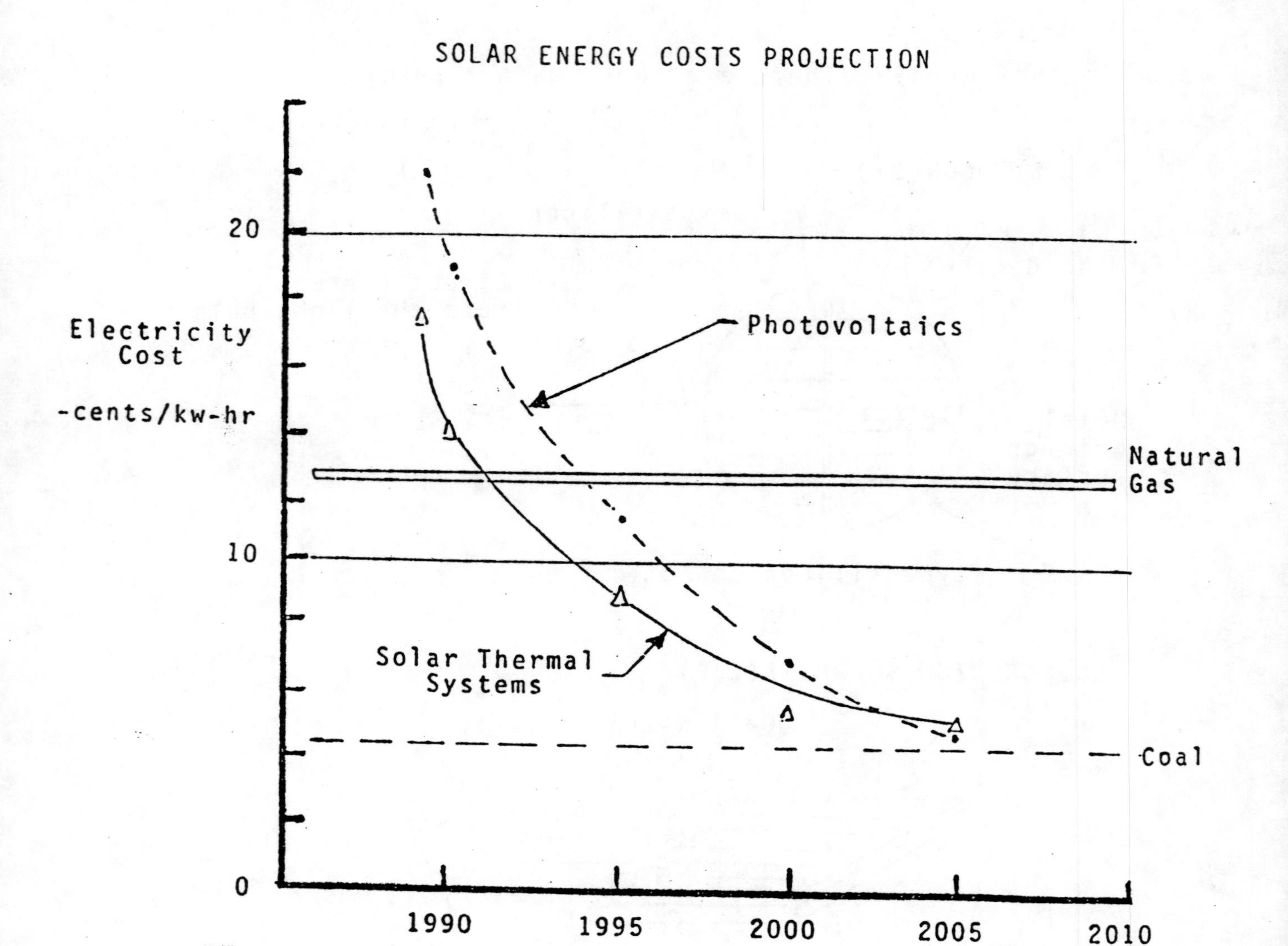

Adapted from Weingberg(1990)

NEW HIGH-EFFICIENCY SOLAR CELL

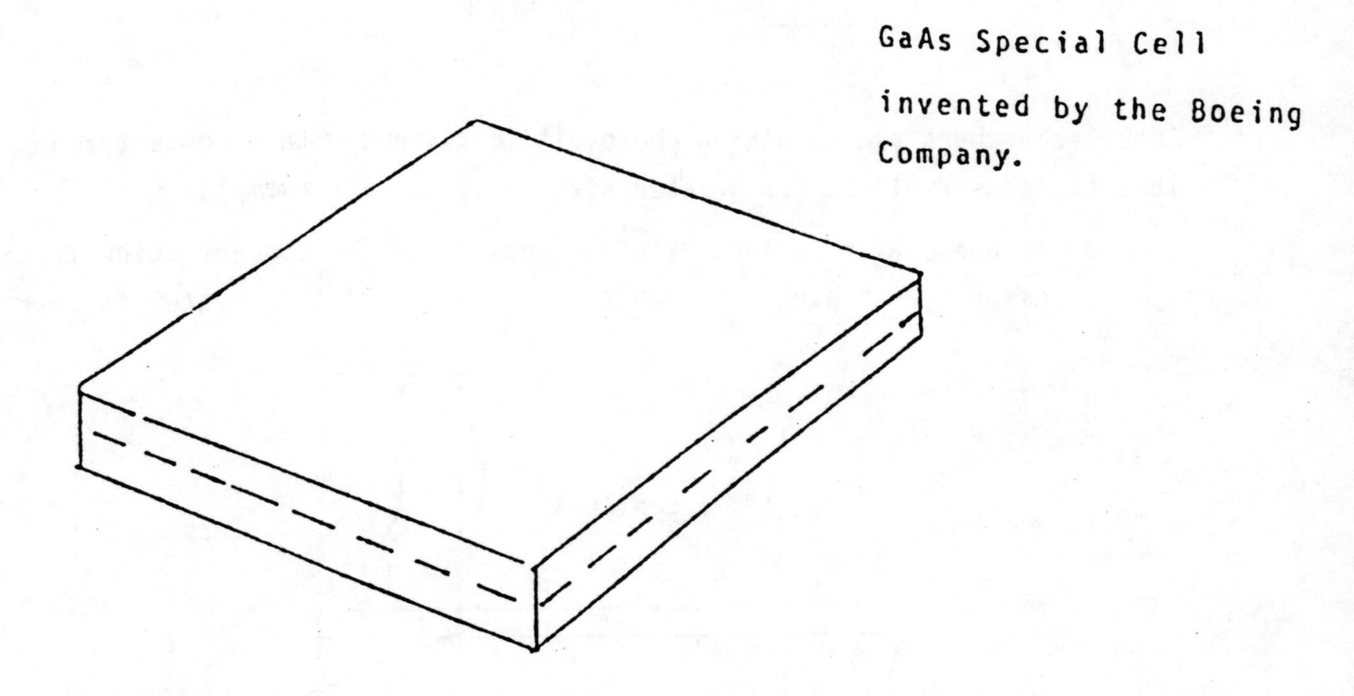

GaAs Special Cell invented by the Boeing Company.

CHARACTERISTICS.

- This cell has managed to obtain an efficiency rating of 37%.
- It is unique in that it uses two materials to catch all the sunlight.
- The materials are :
    - Gallium arsenide
    - Gallium antimonide
- The trick is to catch all the infrared light that passes into the second layer. In most cells made of silicon, the infrared radiation gets lost.
- These solar cells appear to have excellent growth potential.

ADVANCED PHOTOVOLTAICS USING CONCENTRATORS

Some researchers are combining photovoltaic systems with a concentrator lens to focus sunlight for greater efficiencies. For example:

- In one configuration, PV efficiencies of over 20 % have been achieved.
- The concentration factor of this system is over 100.

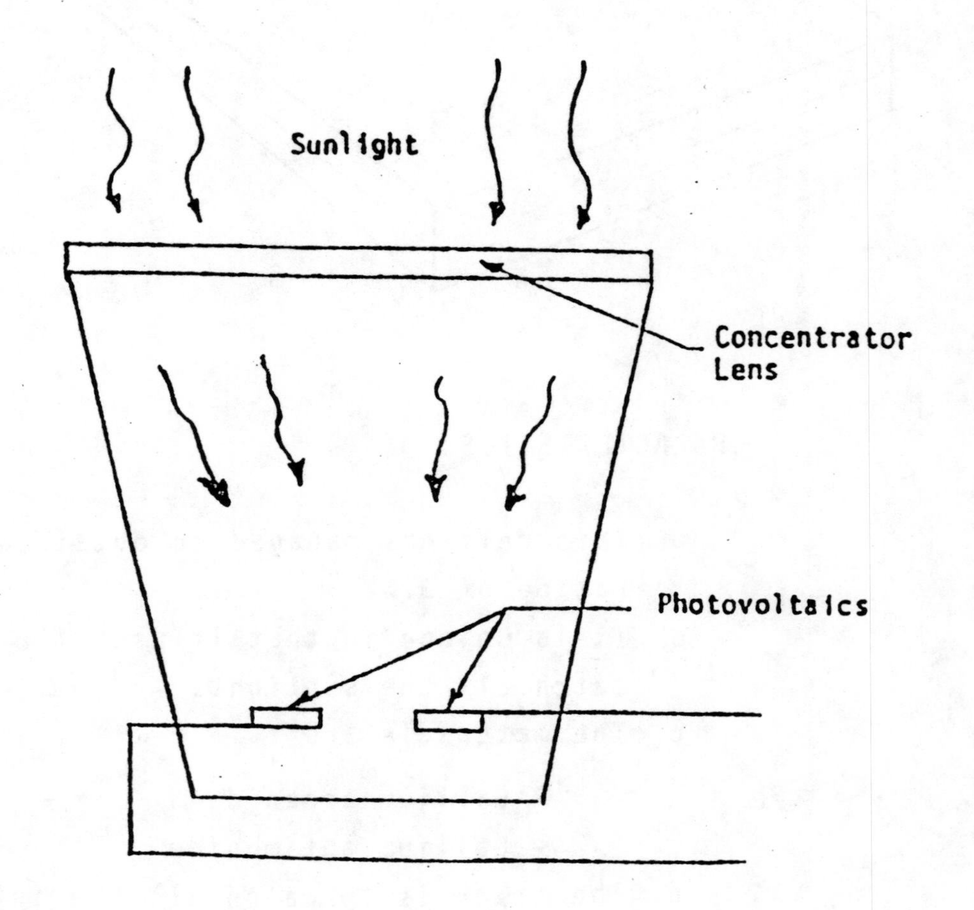

Sketch of the Cross Section of a PV Concentrator. Courtesy of Sandia National Laboratories.

SOLAR CAR

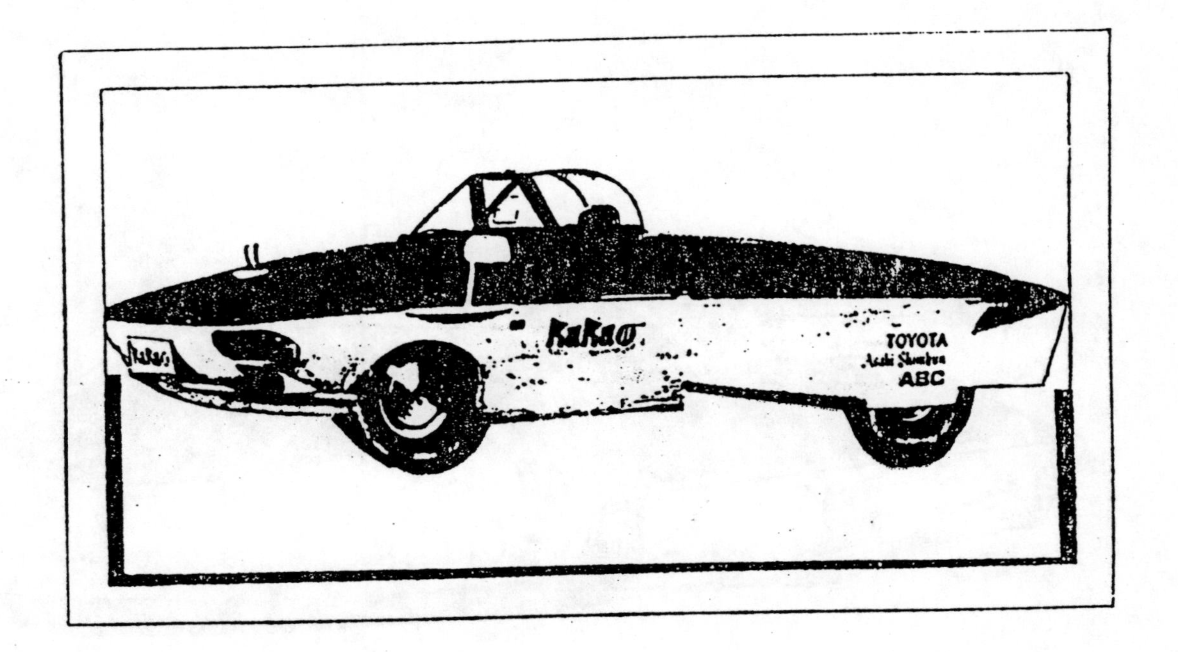

The Toyota Company in Japan has been the first company to obtain a license to operate on Japan's highways.

It has a battery for operation in any kind of weather. However, its top speed is 27.5 miles per hour and it weighs only 463 pounds.

o Source: Popular Science Magazine, March 1991.

RACING ALONG WITH THE SUN

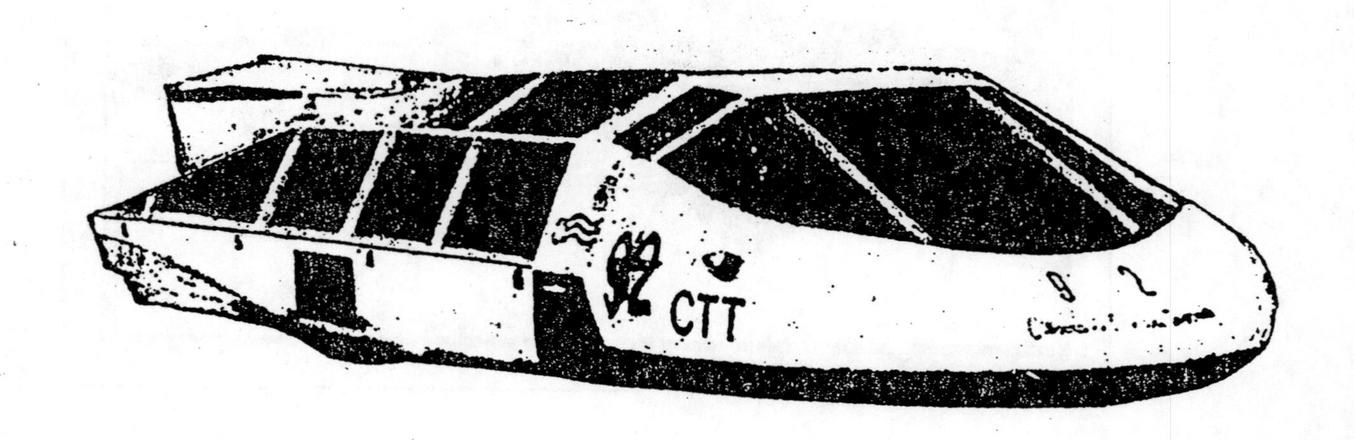

THE SHINING STAR- University of Puerto Rico's entry into the GM Sunrayce.

- It was one of three entries that competed in the 1990 Florida-to- Michigan event in 1990.
- This car was shown at the Epcot Center in Florida.
- The winners in this race went on to compete in the 1990 Solar World Challenge Race in Australia.

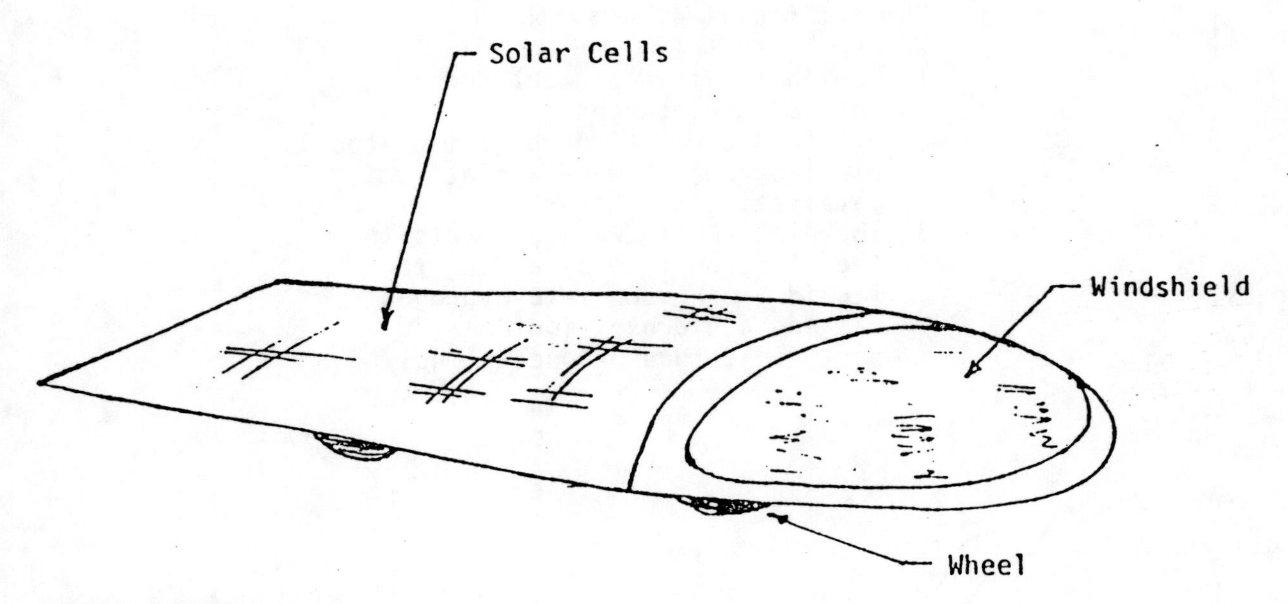

THE SUNRAYCER

A Solar Powered Car

- This car raced 1,950 miles across Australia in 1988.
- Length: 19.7 ft., Width: 6.6 ft., Weight: 360 pounds.
- On a sunny day, the speed was 45 miles per hour.

HOW CAN THE SUN POWER AN AUTOMOBILE?

TWO METHODS

How Artificial Photosynthesis Works

1. Sunlight hits a pigment donor releasing electrons.
2. A second chemical directs the loose electron away toward a platinum catalyst.
3. The platinum(in water) reacts to the electron flow by aiding the liquid break down into hydrogen gas and a hydroxyl ion(charged particle), thus producing fuel.

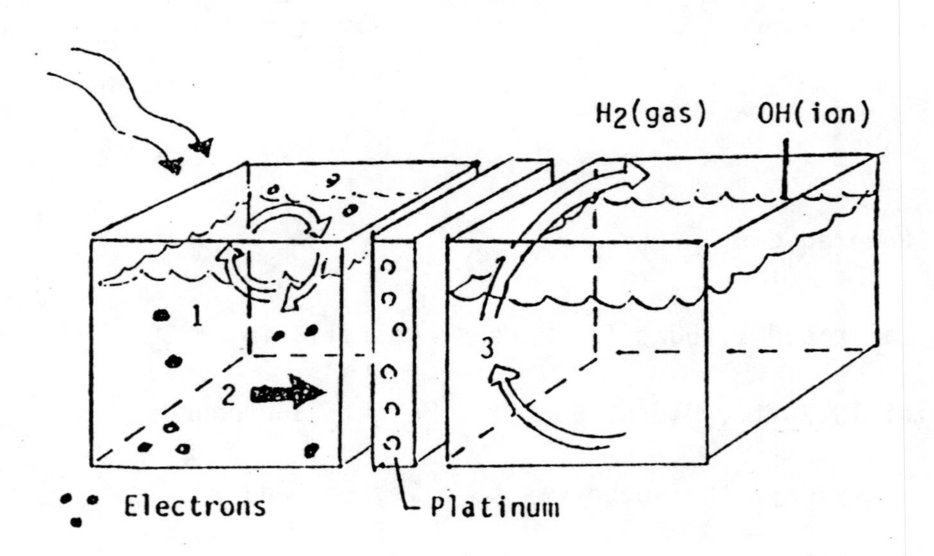

SOURCE: General Motors

ANOTHER METHOD

Point Contact Cell

Point-Contact Cell Can Turn light into Power

1. Sunlight hits silicon atoms,creating loose electrons and "holes"(absence of electrons).
2. A strip across the back of the cell has 73,000 contact points,half positive and half negative. All contacts of each type are connected.
3. Electrons migrate to negative-type points where the silicon has been doped with phosphorus. Holes find positive-type (boron-doped silicon) contacts.
4. When the two areas are connected with a wire, an electrical current will flow.

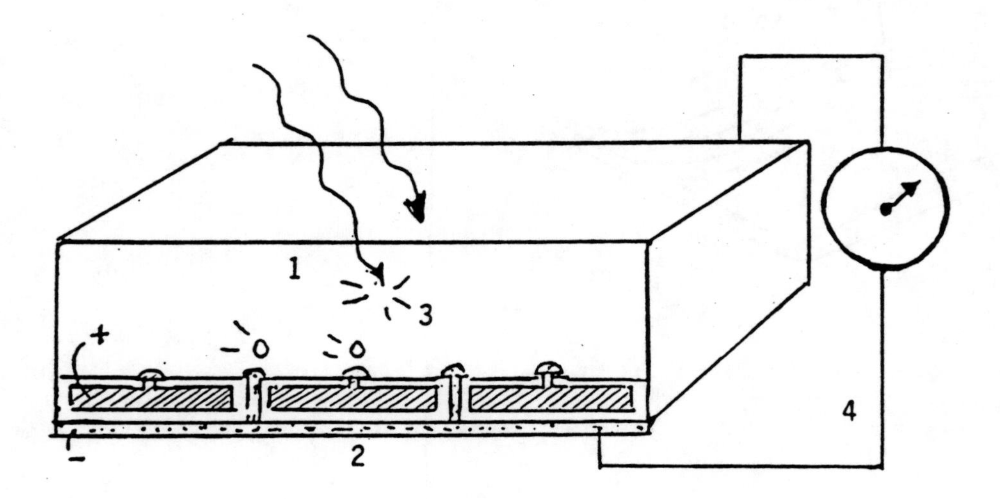

ROOF PHOTOVOLTAICS

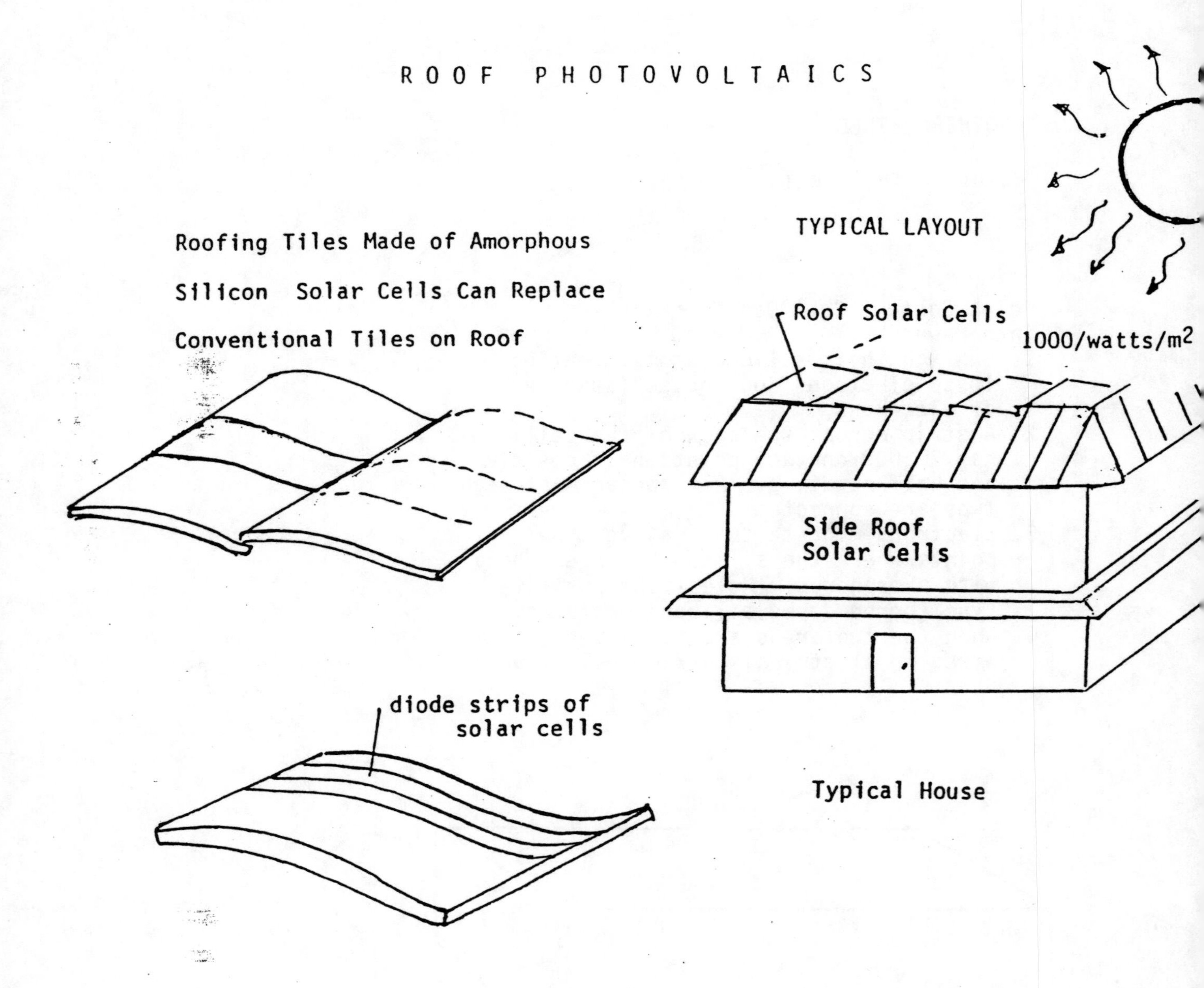

These individual Solar Cell Tiles are made and distributed by Sanyo Electric Company, Ltd., in Japan.

MOBILE RADIO COMMUNICATIONS

AND RV POWER

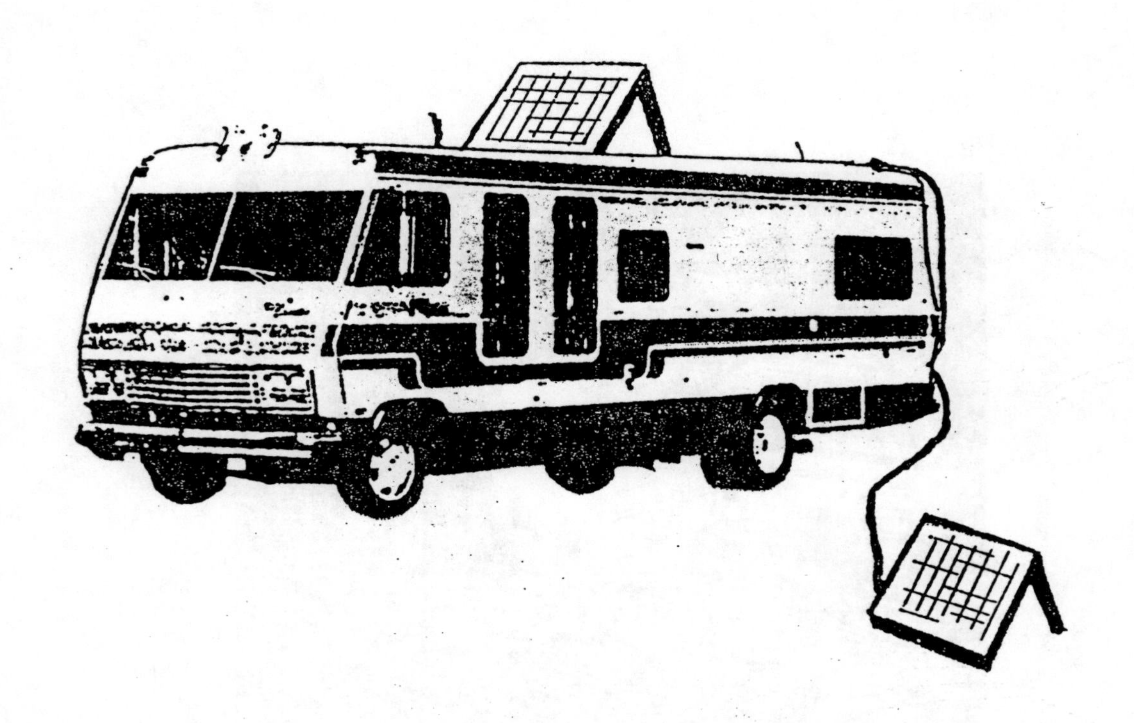

This system is used by vacationers to provide solar power for their TV and other RV appliances.

MOBILE RADIO COMMUNICATIONS USING PHOTOVOLTAICS

This system is used by a western U.S. outdoor club to maintain power for mobile communications

PV MODULES

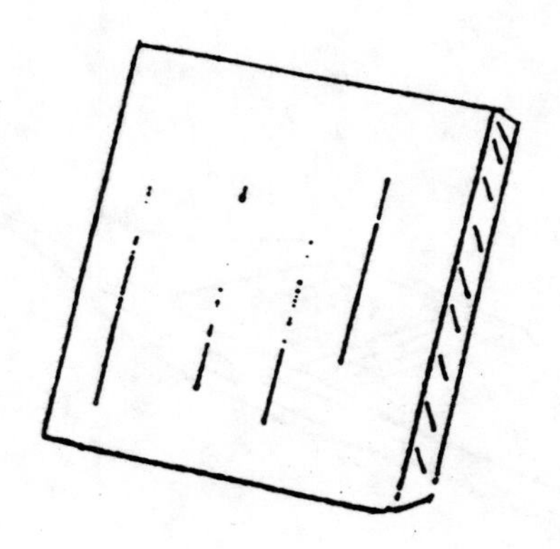

- c Lightweight modules
- o Modules are for recreational or residential use
- o Made in 10 or 20 watt sizes

SOURCE:

SOLAREX CORPORATION
1335 Piccard Drive
Rockville, MD 20850

SOLAR WATCH

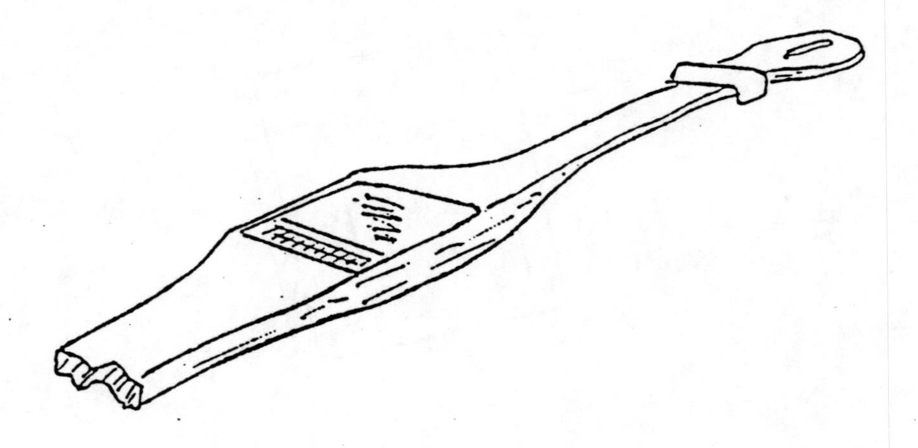

- o Displays continuous digital time
- o Recharges in 2.7 minutes
- o Anti-magnetic
- o Anti-shock

Approx. Cost: $26.00

SOURCE:

RESUMEX PRODUCTS
224 Ardmore Avenue
Ardmore, PA 19003

SOLAR POWERED CALCULATOR

There is at least one computer that uses the sun as a source of energy.

It is available in many store today.

Its characteristics are as follows:

- It is designed for high school level computations.
- It is small and hand-held.
- Primary design is for scientific computations and has 64 functions.
- A manual and a book with solutions is included.

Details: The calculator is available for $12.99.
It is made by the Sharp Co.
It is sold under the Sharp name.

SUN-POWERED RADIO

A radio that works on the energy of sunlight has been developed that can be carried on one's person.

The SOLAR-POWERED STEREO RADIO carries both AM and PM frequencies and is rechargeable.

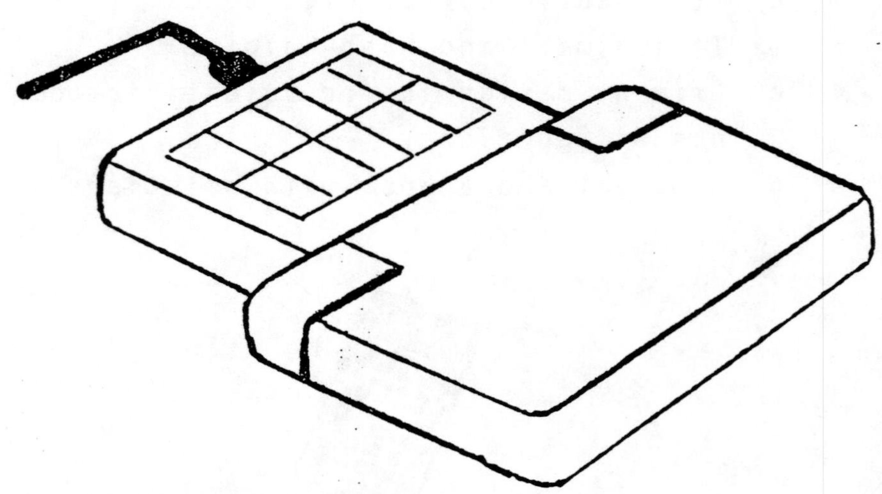

It has the following characteristics:

- It is small with measurements 2.5 x3.5x 0.75 inches.
- The built-in solar panel is charges the nickel-cadmium battery for about 7 hours of listening after sundown.
- It can be recharged electrically when there is no sun.

AVAILABILITY:

KDK World Marketing
Box 931
Upland, California 91785

Source: Popular Science Magazine, October 1991, page 18.

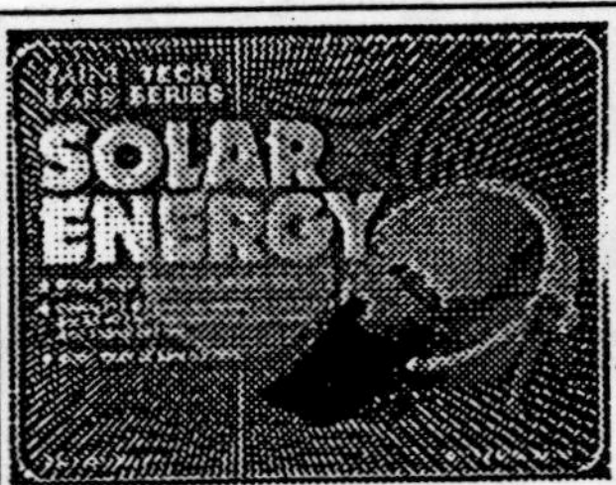

Kit 210 covers experiments with the optics and thermodynamics of solar energy, demonstrating concentration of energy by lenses and reflectors, and conversion of light to thermal (heat) energy. It includes materials and instructions for building and demonstrating a safe working solar furnace!

KIT-210 8 oz. $10.00

Kit 828 demonstrates the direct conversion of light (solar energy) to electrical energy and safely teaches fundamentals of electricity. It contains a small solar module, a working DC motor, and several propellers, spinners, and more.

KIT-828 1 lb. $10.00

Kit 360 contains materials for many of the 150 solar energy experiments described for young persons age 10 and up, and includes optical, thermal and electromagnetic effects. Some of the experiments require only readily available materials not included in the kit. The kit includes both electrical experiments as in Kit 828 and the solar furnace Kit 210. In addition it includes components and materials for building a galvanometer, an electronic temperature sensor, demonstrations of thermal expansion of both gasses and solids, and many other natural physical phenomena. The instruction manual included in this kit is worth the price of the kit!!

KIT-360 3 lb. $35.00

* EMERGENCY RADIO *

SOLAR/DYNAMO/BATTERY

This small radio, built with high quality Sony parts, covers practically every possible way that it could be powered in event of utility failure. It has built-in rechargeable battery, a built-in solar charging panel, back-up rechargeable battery compartment, a receptacle for external solar panel or other DC charger. And if all else fails, it has a built-in hand crank DC generator that you can use to recharge it. With all that, it is also a good quality AM/FM radio.

RA-1 2 lb. $35.00

Kit 689 features more than 24 pieces with 8 PV cells that can be easily wired in any configuration by using jumper bars provided. Kit even includes tools for assembly of the projects! Informative booklet details the whys and hows of solar energy and explains how to interconnect the PV cells to provide different voltages and currents. Kit includes motor and fan with a stand, PV cells, jumpers, tools and wiring. An excellent kit!

KIT 689 1 lb. $19.95

SOLAR SENSOR
10909 Hayvenhurst Avenue
Granada Hills, CA 91344

Chapter 4

SOLAR THERMAL SYSTEMS

The most popular use of solar energy is to convert it directly to heat and use the heat to run various systems. Applications are electrical energy, home heating and cooling.

There are many aspects to the use of thermal energy. Some are shown in this section.

THERMAL SYSTEMS

SOLAR POWER ONE

Large Desert Solar Receiver Power Plant

This solar electric power plant is typical of the large solar systems built in the United States and Europe.

Specifics are:

- o It has 1818 mirrors that reflect the sun's energy to the receiver in the central tower.
- o The mirrors are large, weather durable and computer controlled.
- o This particular system, located near Barstow, California was closed because of economic considerations in 1989.

SPACE HEATING

Innovations in Space Heating

Perforated Siding is Used on the Sun-side of the House.

- o Prefabricated panels are 2 foot x 2 foot in size and they are built to hook together.
- o Typically, they can provide 20-30% of the home heating costs.
- o Amount required: 1 square foot of wall space for every 100 square feet of living space.
- o They are asthetic in appearance on the outside of the house.
- o Prices are approximately $5-6 per square foot.
- o Available from Energy Research*

*Mr. J. P. Trainor, Solar Engr. & Contracting, Vol 1.,No. 11, pp. 19-20, November 1982.

SOLAR HOT WATER

What are solar hot water system costs? ?

Many types of hot water systems exist for solar energy applications.

In a typical, simple solar system, the costs are low.

As an example:

Burton's Sundulator Collectors.

Roof-top installation price: $4.50 per square foot.

Available from the following source:

Mr. John Dioguari

BEST COMPANY

243 Wyandanch Avenue

North Babylon, New York 11704

Telephone: 516-643-6660

The total cost of complete solar hot water systems is $3000-$3200, although the net cost, after tax credits, may be 50% or less.

S O L A R   H E A T I N G

HOW MUCH HOT WATER DOES A TYPICAL HOME NEED ? ?

- o Determining the optimum size of a hot water heater for home use can be complicated.
- o Contractor experience to-date says -
  - Always install a 120 gallon water storage tank and approximately 100 square feet of collector.
  - This produces all the hot water necessary for use in a typical house located in the Southern United States while being partially adequate in the North.
  - The lengthy and complicated calulations are thereby avoided.

SOLAR HOT WATER HEATERS

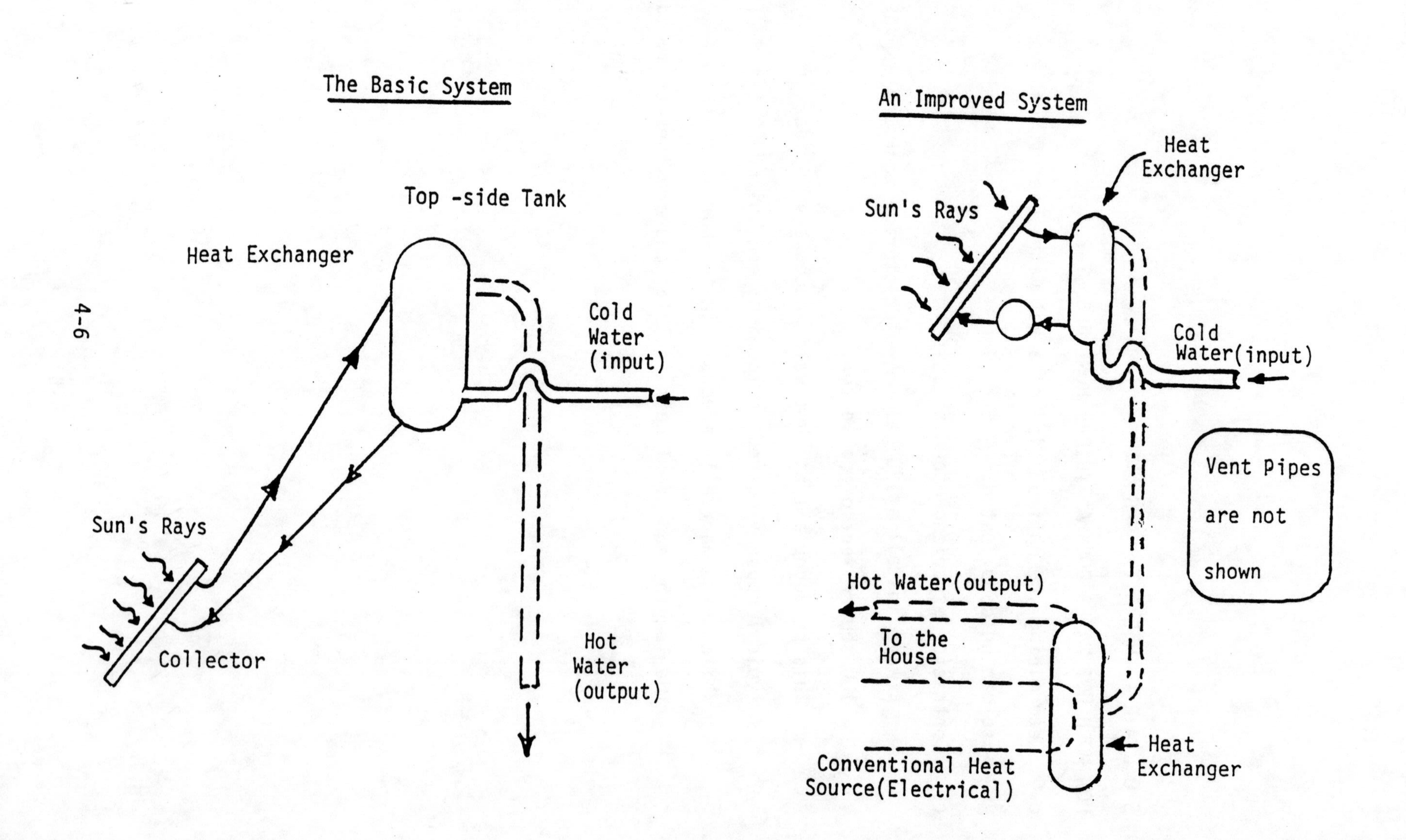

INSULATION FOR HOT WATER HEATERS

To save energy:

- o Insulate the pipes to a hot water system with molded polyethylene pipe sleeves. They are easy to install.
- o Lower the operating temperature of the thermostat will also save on the monthly utility bill. For example:
  - Lowering the setting from 160° F to 140° F will save about 7 %.
  - Lowering the setting from 160° F to 120° F should save some 20%.
- o Insulate the hot water tank.
  - For an expenditure of $25.00(large tank) to only $15.00 (small tank), a savings of some $25.00/year on the utility bill may be achieved.
  - The Do-It-Yourself installation kit pays for itself in a year.

NON-CONVECTING SOLAR PONDS

Black-bottomed ponds are used to store heat from the sum. Typical ponds are 1 meter deep.
Salt is added to supress the convection.
The solar energy is trapped due to the gradient effects within the liquid.

Temperatures approaching the boiling point of water have been recorded.

The energy saved from this source can be significant.
Low maintenance is required.

SOLAR PONDS

Cross Section Drawing of an Advanced Solar Pond System

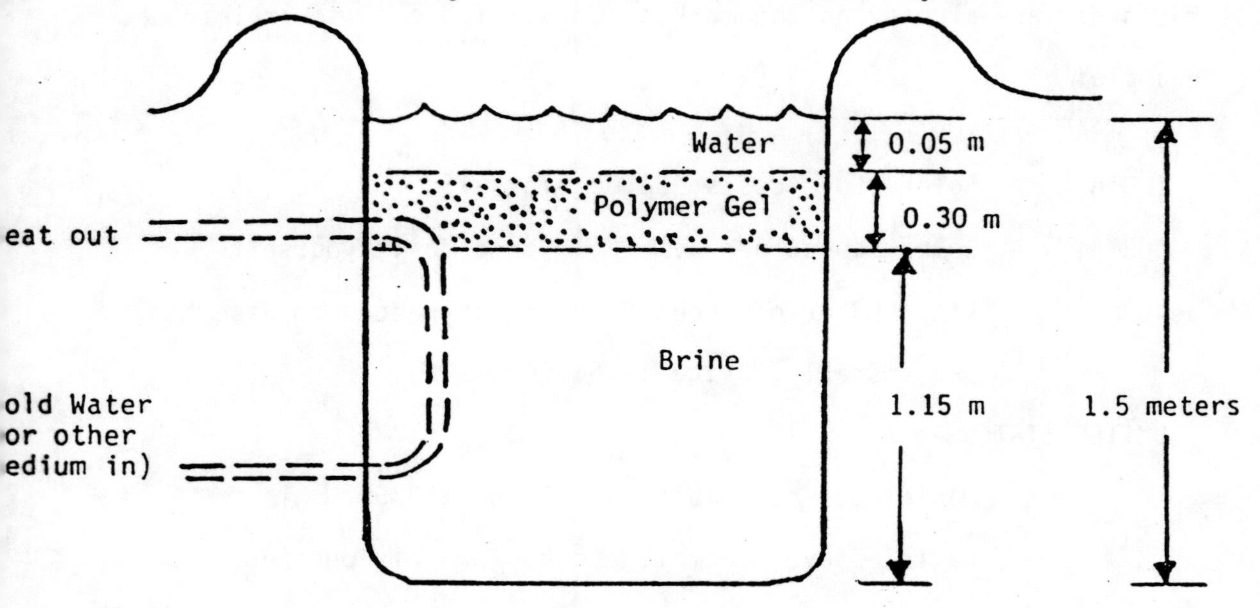

The area of the pond may be wide. This type of improved pond system for recovering the sun's heat has recently been developed.

Some 200 polymer gels have been researched: Four have been found to be best.

For further information: Mr. E. S Wilkins, Dept of Chemistry and Engineering, University of New Mexico, Albuquerque, NM 87131.

SPACE HEATING

SOLAR STOVES FOR COOKING AND HEATING

Solar stoves are already on the market and available in several types. For example:

INDIA - Solar stove costs $8.00 in 1975.

BURMA - Solar stoves are available in a do-it-yourself kit. Labor of about 8 hours is needed to assemble and install. Price is $1.00 (1975).

UNITED STATES -

Daniel Stove, Plastic $2.00 (1975).

Davis Stove, Metal. Cooks for only one person. It needs to be refocused every 20 minutes in order to follow the sun. Price is appr. $14.00 (1975).

OTHER SOLAR HEAT APPLICATIONS

Salt Water Distillation

A useful method for solar energy. As early as 1872 , a 4700 square meter system was successfully built in Chile.

What are the costs?

The costs , of course , vary with the type of system. Typical examples are:

Size Costs: 0.1 dollars per square meter operation costs.

Capacity costs: Costs of obtaining water by means of these systems varies from $ 1 per gallon of water (daily).
to $ 6 per gallon of water (daily).

MANY TYPES OF DIFFERENT COLLECTION SYSTEMS EXIST. PRICES WILL VARY WITH THE SIZE AND CAPACITY OF THE DEVICE.

For additional information: See the references. Kreider and Kreith.

SOLAR POWER COLLECTOR

TYPICAL SOLAR COLLECTOR FOR PROCESS HEAT

Maximum process heat capability: 650 °F.

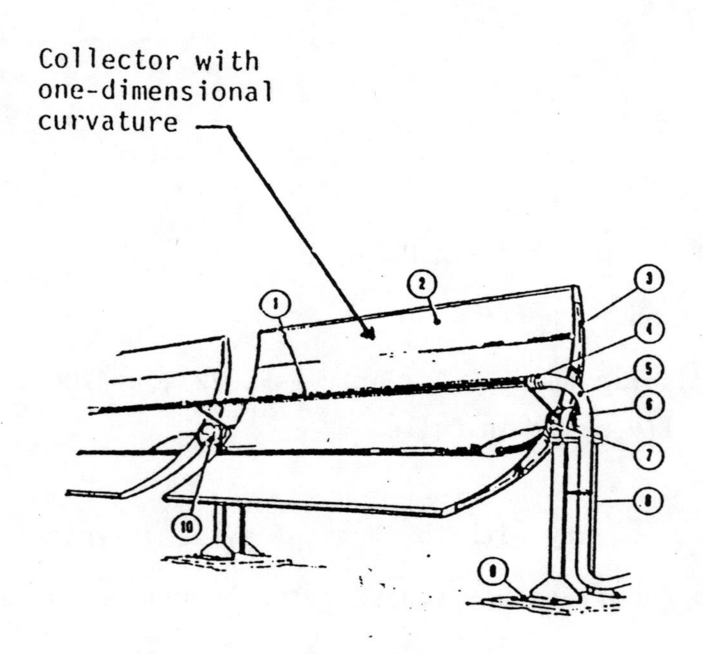

Solar Kinetics' rigid monocoque* mirror and no-lash hydraulic tracking* along with proven engineering concepts provide long life and low cost. Features of the system are:

(1) The black chrome plated steel receiver tube is surrounded by a dry air annulus protected by Pyrex* glass tubing. Focus is adjustable during installation.

(2) A precisely constructed mirror surface is covered with metallized acrylic film or glass combining weather resistance and high reflectivity.

(3) The parabolic contour is N/C machine generated for an accurate focus.

(4) The thermal expansion bellows allows for expansion of the receiver assembly and maintains a sealed, dry environment in the annulus.

(5) An insulated stainless steel flex hose allows rotation of the collector with unrestricted flow.

(6) Self aligning sealed ball bearings absorb structural loads maintaining collector motion without binding.

(7) A steel flange carries torsional loads into the collector structure. Allows mirror installation with 10 bolts.

(8) The steel support pylon is galvanized for corrosion protection.

(9) Mounting studs are a standard pattern for each collector.

(10) This load bearing joint* protects the collector structure from strains induced by misalignment from foundation shifts.

SPECIFICATIONS	T-2100
MODULE WIDTH	22'
MODULE LENGTH	20 FT
MIRROR WIDTH	21'
SOLAR AREA	410 FT²
REFLECTANCE	0.84/.96
MAX VERT HEIGHT	24'
ROTATION AXIS HT	13'
TRACKING ANGLE	270°
STOW ANGLE	HORIZONTAL
SYSTEM WEIGHT	5.0 LB/FT²
RECEIVER TUBE	4.5" OD
ANNULUS MEDIUM	DRY AIR
SELECTIVE SURFACE	BLACK CHROME
ABSORPTIVITY	0.94-0.97
EMISSIVITY	0.18 @ 500°F
RECEIVER COVER	PYREX GLASS
MAX OPER TEMP	650 F
MAX OPER PRESS	250 PSI

For further information: Solar Kinetics, Inc., 10635 King William Drive, Dallas, Texas. Tel: (214)-556-2376.

MEDIUM-TEMPERATURE

COLLECTOR

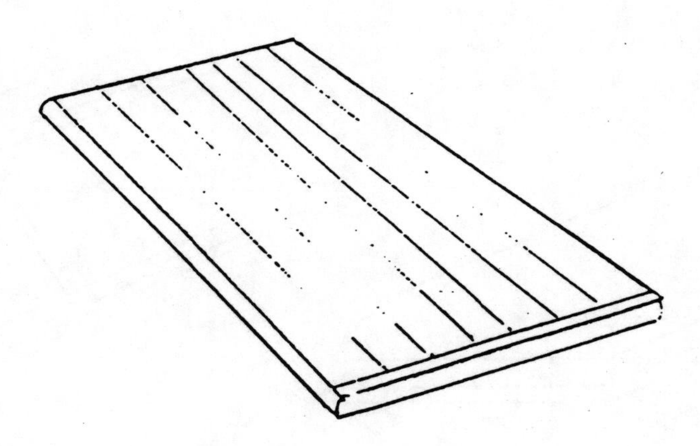

This three-inch deep solar collector is for domestic, commercial or industrial applications.

Features:

- o The Trimline collector has an advanced absorber plate with an integral selective surface, a reinforced aluminum housing and a modular quick-set mounting surface.
- o Absorber plate consists of Al fins cold-welded under high pressure. This is to reduce corrosion.
- o Selective surface characteristics are:
  - Minimum absorptivity = 0.95
  - Maximum emmissivity = 0.15

SOURCE:

Solar Industries, Inc.
2300 Highway 34
Manasquan, New Jersey 08736

SKYLIGHTS SAVE ENERGY

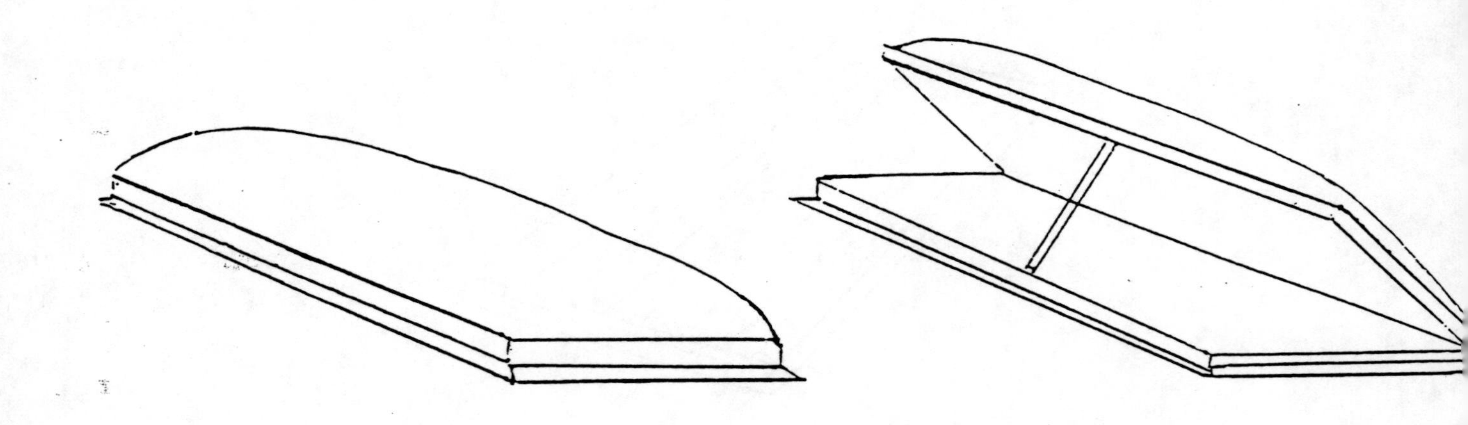

Insulation value is five times that of wood.

SOURCES:

a. Fox Plastics Corp., 8300 Dayton Road
   Dayton, Ohio 45324
   800-233-FOXX
b. Tub-Master Corp., 413 Virginia Drive,
   Orlando, Florida 32803
   800-821-1154
c. Velux-America, P.O. Box 3208
   Greenwood, South Carolina 29648

SOLAR COOLING

Since air conditioning is widely used in the United States, solar cooling systems must be compatible with them.

In 1977 for example:

- Central air conditioning was used in 53% of all residential construction.
- Number of Air Conditioners Sold:

  Central Air Conditioners: 2.6 million.

  Room Air Conditioners: 2.8 million.

Many specific systems exist.

The rating(capacity) of the system is measured in "tons" which refers to the amount of ice that the system could theoretically form.

For solar systems, the amount of storage capacity built into the system is all important.

SOLAR COOLING

Space can be cooled by the use of solar energy.

How can this be done? Two systems have been proposed.

A. Absorption Chillers:

As an example, solar heated hot water can be designed to drive an absorption chiller.

B. Rankine Compression Refrigerator Cycle:

In this concept, a Rankine cycle turbine is used to drive a conventional compression type chiller. Further work needs to be done to prove the practicality of small turbines and increased efficiencies above 12 %.

Frequently, a computer simulation of this complex system is needed to evaluate the effectiveness of this proposed cooling systems.

HOUSE COOLING

Solar energy can pump considerable heat into a building or home directly through the attic.

The installation of a solar energy heat barrier is a relatively simple matter and can be installed in a short time.

During mildly hot weather, this may reduce the heat sufficiently so that an air conditioner need not be turned on - hence, energy is saved.

SOLAR HEAT BARRIERS

- O Aluminum foils are good for heat reflection although other types may suffice.
- O Heavy duty foils or foils reinforced with fibers or paper work best.
- O Foils are stapled into position on the bottom of the attic rafters.
- O Have vents and try to channel the trapped heat up to a ridge vent.
- O Heavy duty foils are available for this purpose in 4 and 5 foot rolls.

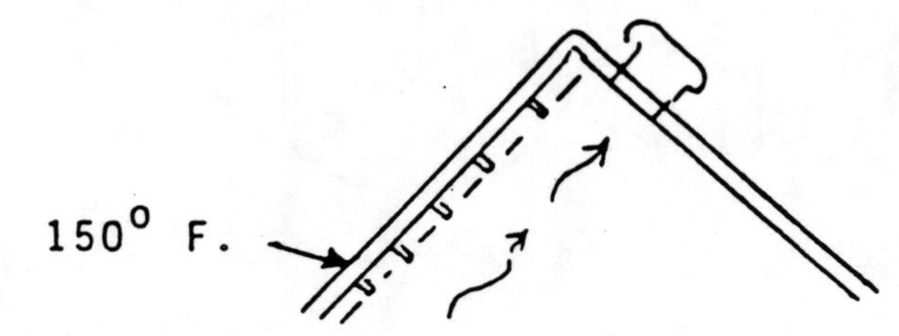

SOLAR LIGHTING

Toronto, Canada Building

The concentrators follow the sun. When cloudy the electric light bulbs give continuous light.

The light is piped to all parts of the building.

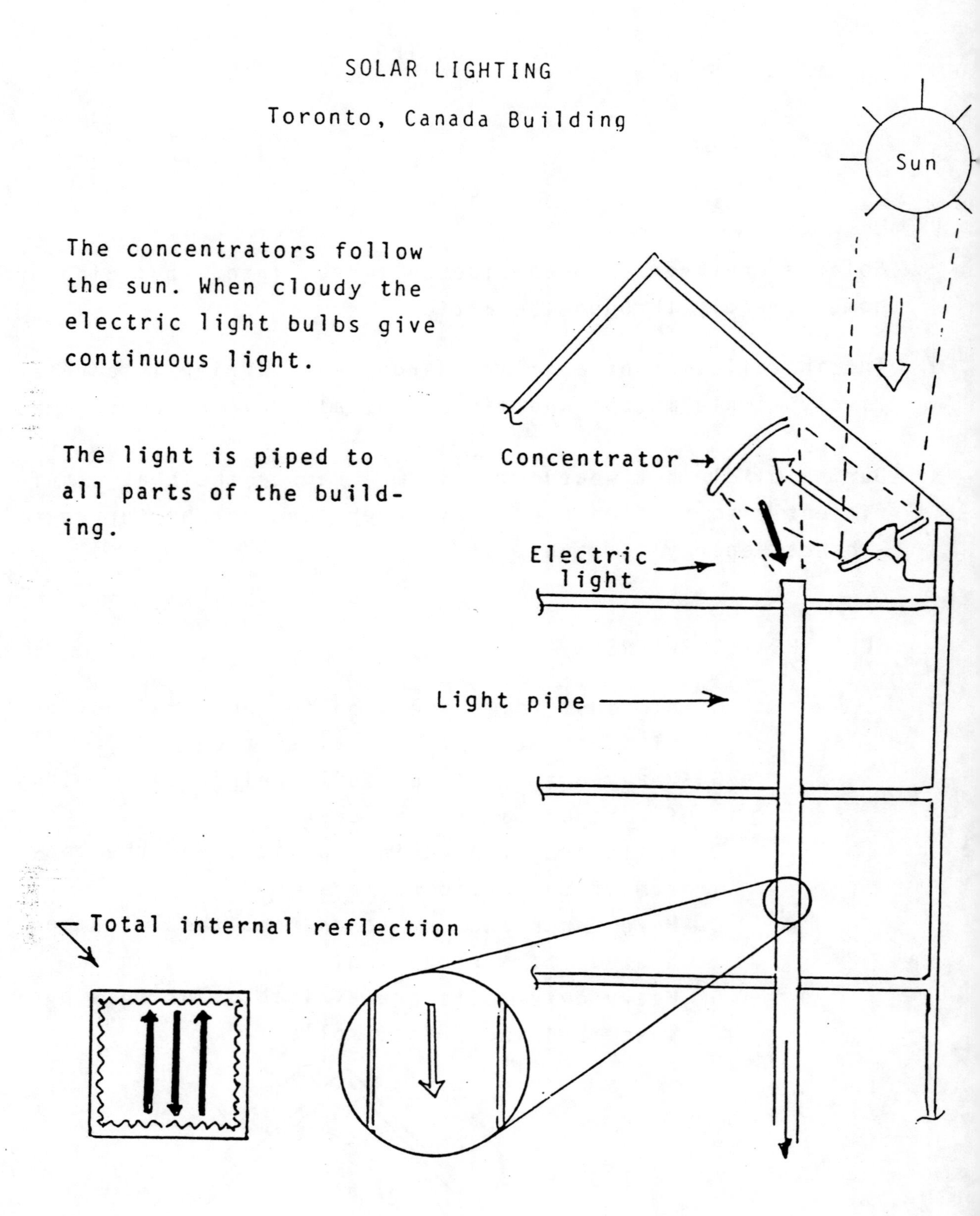

METHODS OF SOLAR ENERGY COLLECTION

In addition to the normal methods of straight-foreward collection of solar energy, such as the trapping of energy through windows by direct absorption, it is possible to concentrate solar energy in many ways.

Some typical methods are summarized below:

Flat plate reflector that is one of many that focuses into a receiver-

a. Stationary

b. Solar tracking(follows the sun via computer inputs).

One-Dimensional Concentrator(a groove).

a. Stationary

b. Solar Tracking.

Two-Dimensional Concentrator(dish)

a. Stationary

b. Solar Tracking.

Sketches of the concentrators are shown.

METHODS OF SOLAR COLLECTION

Three methods of solar energy collection through concentration are given on the previous page. The following generalizations can be made:

- Solar tracking systems are more efficient than stationary systems.
- Solar tracking systems are more complex and hence more expensive.
- Solar tracking systems are usually less relaible than stationary systems.

Additional systems exist.

For example, in terrestrial, one-dimensional concentrators, solar collectors have been successfully used in conjunction with photovoltaic systems.

The reflectors enhance the energy collected on the surface of the PV cell.

Typically, two to three times the energy can be collected in this type of configuration.

Other combinationa of various PV and solar collectors have been investigated. Many engineering articles exist.

GIANT DESERT AREAS FOR SOLAR POWER GENERATION

Giant solar farms have been developed in the American U.S. southwest deserts to produce electricity with sunlight.

This is a tremendous advance in the program to produce large scale electricity for counties and cities.

Some important aspects of these plants are the following:

- Over 275 megawatts of electricity is now being delivered to the Mojave Desert region.
- Over nine plants have been built using the principle of parabolic reflection of the sun's energy in solar concentrators.
- Due to large scale production efficiencies, the newer plants produce electricity cheaper than nuclear power.
- According to LUZ ENGINEERING CO., builder of the 80 megawatt plants, each plant shrinks the carbon dioxide emissions by 325 million pounds annually when compared to oil-fired plants.

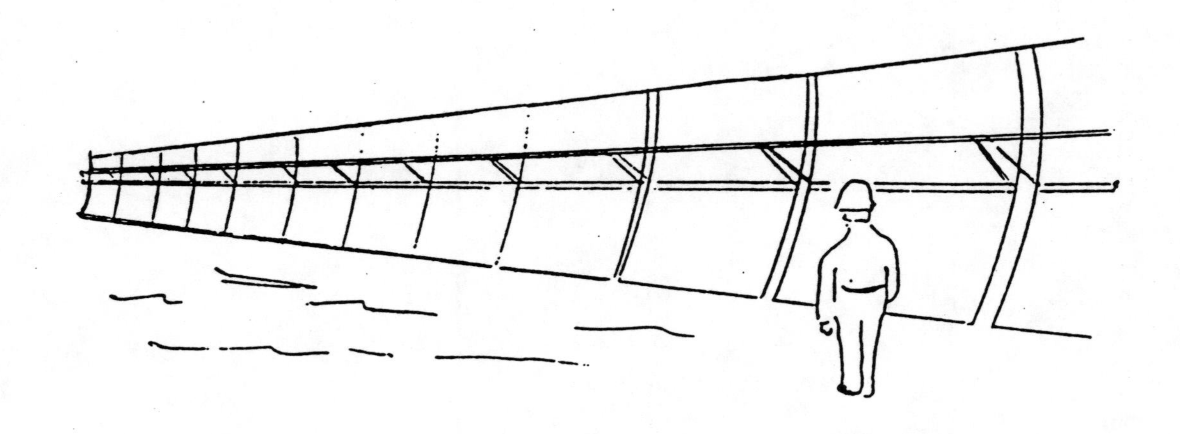

SOLAR HEATING

-Solar Greenhouses

A word should be said about solar greenhouse, perhaps the most popular use of solar energy today, in the United States as well as foreign countries.

Characteristics are:

- They are used to grow fruits, vegetables and flowers.
- Geometrically, they may be divided into three general types, depending on the size and relationship to the house. See sketches.
- Solar materials of three basic types are used to transmit and trap the solar energy:
  a. Glass
  b. Plastic
  c. Fiberglass

SOLAR HEATING

-Solar Greenhouses(Cont.)

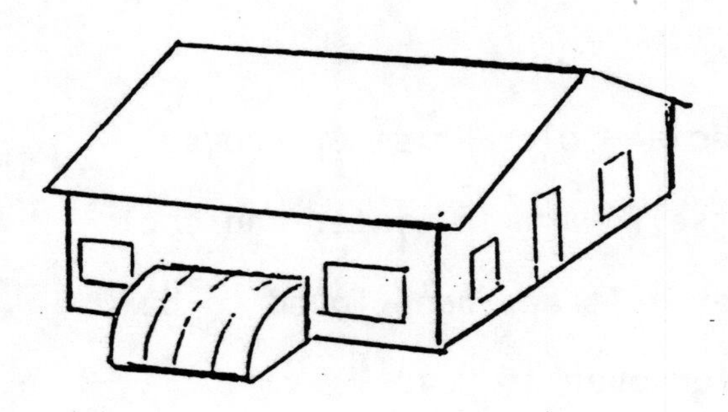

Small Additions

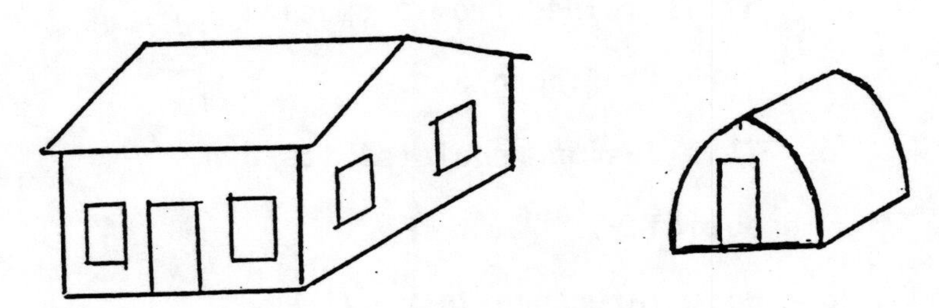

Small Independent Structures

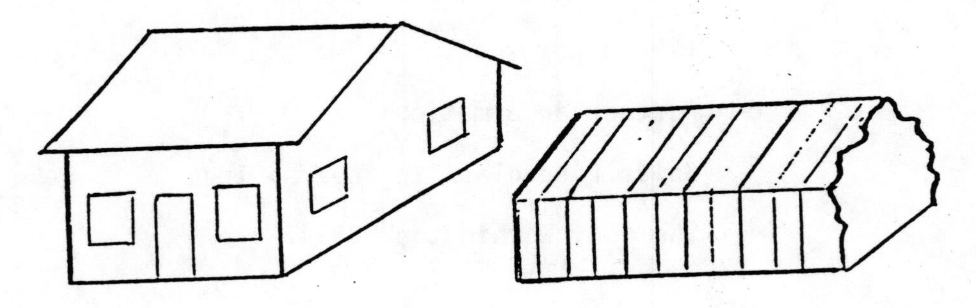

Large Independent Structures

Further information on a small independent greenhouse is given on the following page.

S O L A R G R E E N H O U S E

-Special Design

This special Solar Prism Greehouse keeps itself warm in winter and cool in summer. It has been found to be useful for over 18 years.

Some features include:

- o It is molded from a special formula fiberglass.
- o This design regulates its own humidity.
- o No maintenance is required.
- o Heating costs are only $25.00 (1989) per year.
- o Further information:
  Solar Fiberglass Engineers, Inc.
  Sheldon, Washington 98584.

BUILD YOUR OWN GREENHOUSE

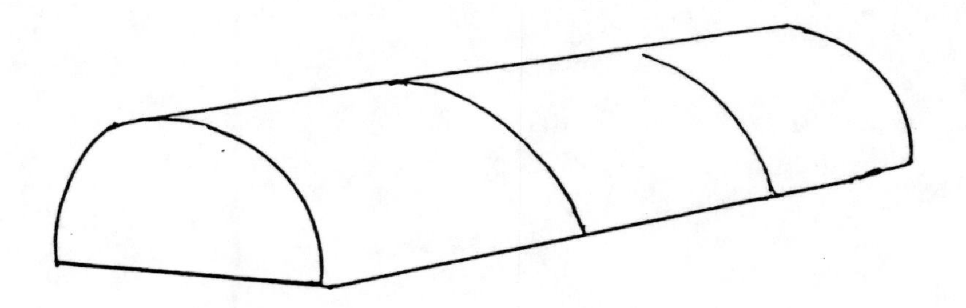

Using PVC piping, you can build your own storage facility that can be used for a greenhouse.

Typical uses are:

- o Greenhouse. Increase the growing season for your flowers or crops.
- o Use as a solarium.
- o Storage. Use to store various items including wood.
- O Use as a garage for your tools or car.

FOR THE PLANS:

Write to: RTB Enterprises, Dept. HM
P.O. Box 3338
Auburn, Maine 04212

SOURCE: Home Mechanics, June 1991

BLANK PAGE

MEDIUM TEMPERATURE
SOLAR COLLECTORS

System shown with optional self-contained Photovoltaic Power Pak™

- These collectors are designed for the medium temperature range, namely 120 °F. to 200 °F.
- Useful for hot water, heating and air conditioning units.
- High quality.

SOURCE:

AMERICAN SOLAR KING CORPORATION
7200 Imperial Drive, P.O. Drawer 7399
Waco, Texas 76710

SOLAR COLLECTOR

This evacuated heat pipe solar collector works in ALL types of weather- with nearly twice the efficiency of conventional flat plate collectors.

SOURCE:

Maine Solar Technics
360 Center Street
Bangor, ME 04401

207-947-2909

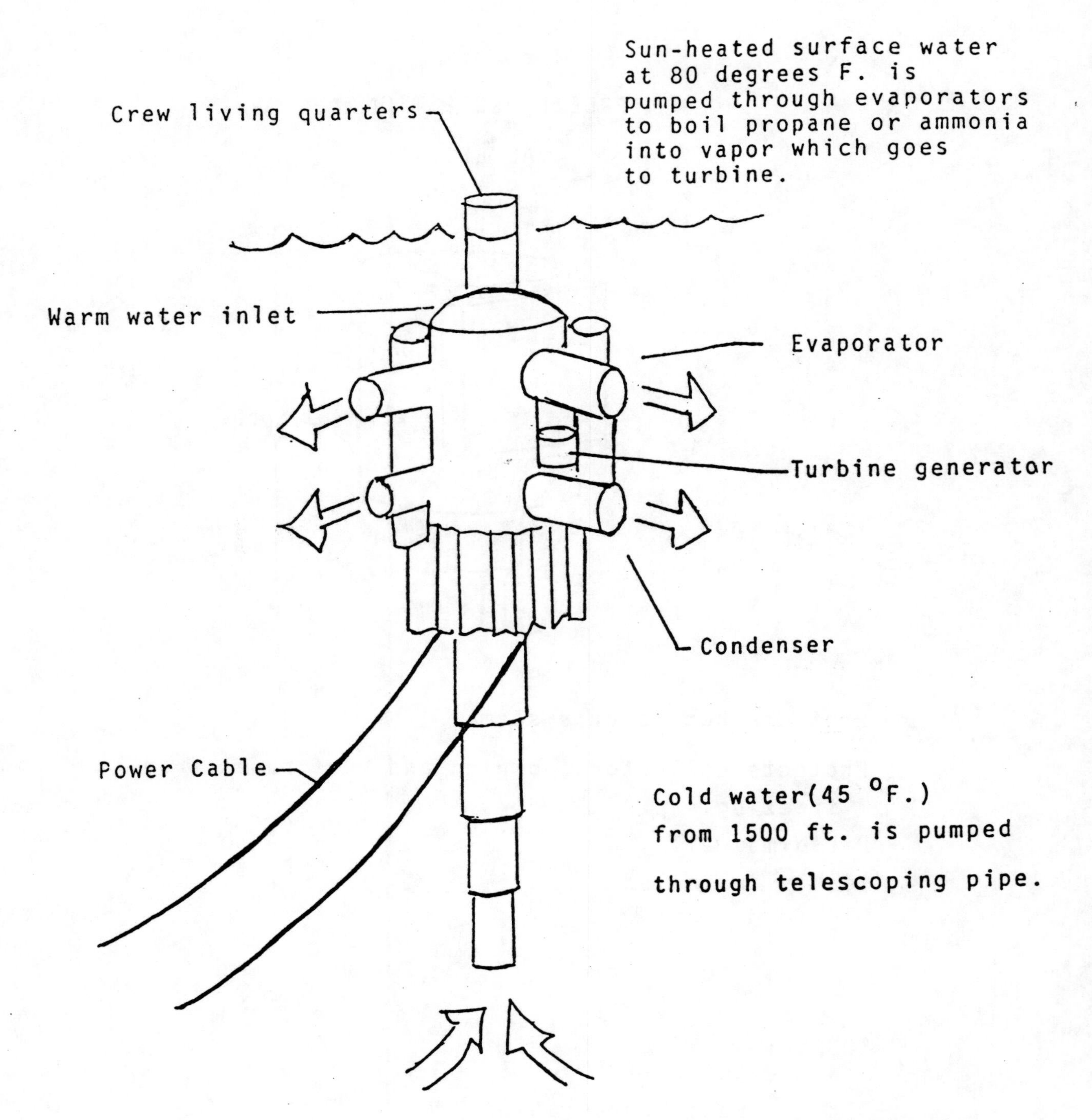

Source: Lockheed

FREEZE PROTECTION

VALVES

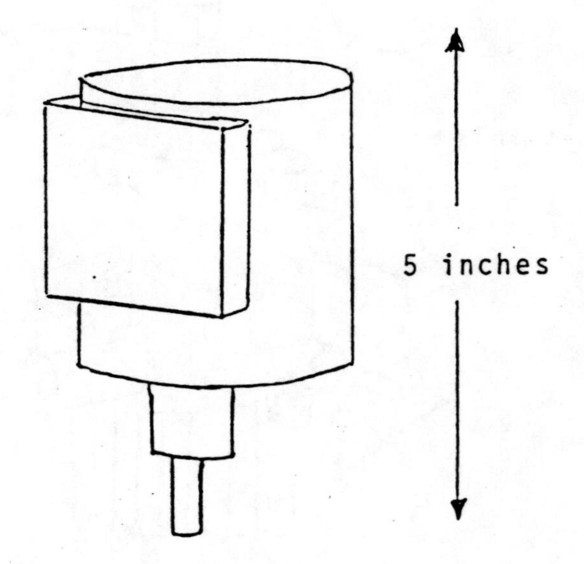

Protects hot water systems.

Protects collectors, piping and heat pumps from freezing.

Performs well

Easily installed

SOURCE:

Eaton Corporation
Controls Division
191 East North Ave.
Carol Stream, IL 60188
Tel: 312-260-3034

FUTURE OF SOLAR COLLECTION

Solar Converters of High Efficiency

Recent studies have shown that there is an increased possibility of developing greatly more efficient solar energy collection systems. Specifically:

- o Natural systems have been found to convert solar energy to heat with much greater efficiency than man-made systems.
- o Specifically, the hair of a polar bear, although white, converts energy at greater than 95 percent.
- o This unique trapping system within the hair is transparent, not black.
- o The solar energy is reflected down the hair in such a way that little energy is reflected. Hence, it is trapped on the bear's skin and keeps it warm.
- o Some experiments that have applied this technique to flat solar arrays have shown increased efficiencies.
- o Application of this technique to other solar energy systems may be possible.

For further detail, see:

Solar Polar Bears, Scientific American, Volume 258, No. 3, page 24, March 1988.

PURIFYING WATER WITH SUNSHINE

A water purification plant has been built that uses sunlight as the energy source. The layout of the plant is shown roughly below in the sketch.

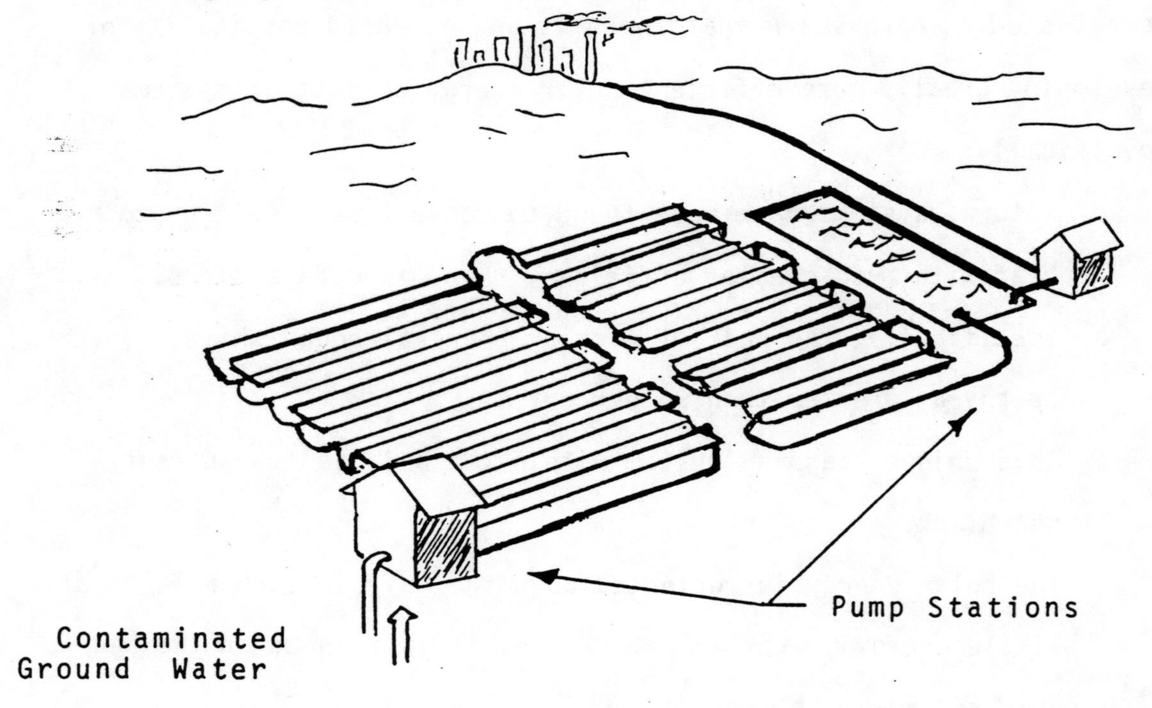

This plant is supervised by the Solar Energy Research Institute of Golden, Colorado. One of the purposes of the plant is to clean up spilled dyes, spilled fuels, pesticides and other organic wastes.

The solar energy is capable of heating up the gases to approximately 1832 ^{O}F. with intense heat and UV light.

Source: Church, Vernon M., Purifying Water with Sunshine, Popular Science, August 1991, page 26.

DOMESTIC SOLAR HEATING USING ROCKS

Rocks can be used to partially or totally heat houses or businesses. An example is given.

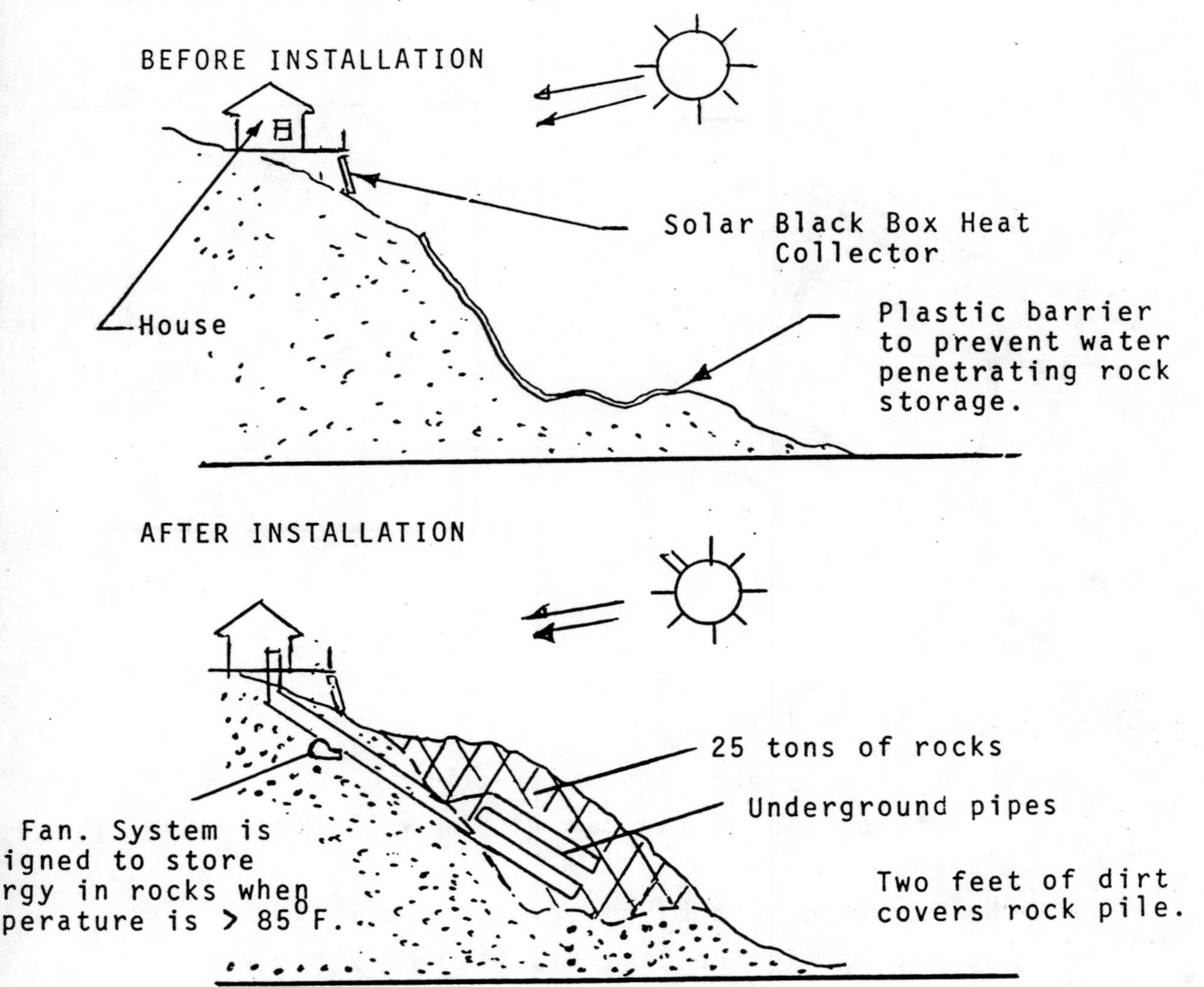

Twenty-five tons of rocks were successfully used as part of a domestic solar heating system. It was installed in Port Ludlow, Washington and has been used for a decade.

Source: Stocker, H., Domestic Solar Heating Using Rocks, SYSTEMS CO., Inc., *2nd* Edition, 1993.

SOLAR CAR VENTILATOR

This system has been designed to keep your car cooler when it has been parked all day.

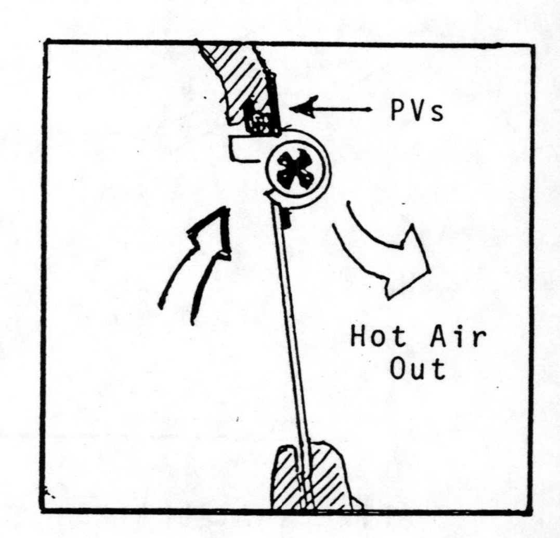

Specifics are the following:

- The system is entirely solar-powered.
- It fits on top of the car window with the window closed.
- The fan is driven by photovoltaic cells.
- The hot air in the car is continually replaced.

AVAILABILITY: Send $1.00(1991) for system details and prices. to:

LIVE NEWSLETTER Issue 91,
P.O. Box 2010
Flòra Park, New York 11002

SOLAR LIGHTING SYSTEMS

Lighting systems have been developed for use in lighting homes, tall buildings and underwater or ground systems.

One system is installed in a high-rise building in Tokyo, Japan. It is made by a Japanese company. The plastic enclosed solar cells are installed on top of the building.

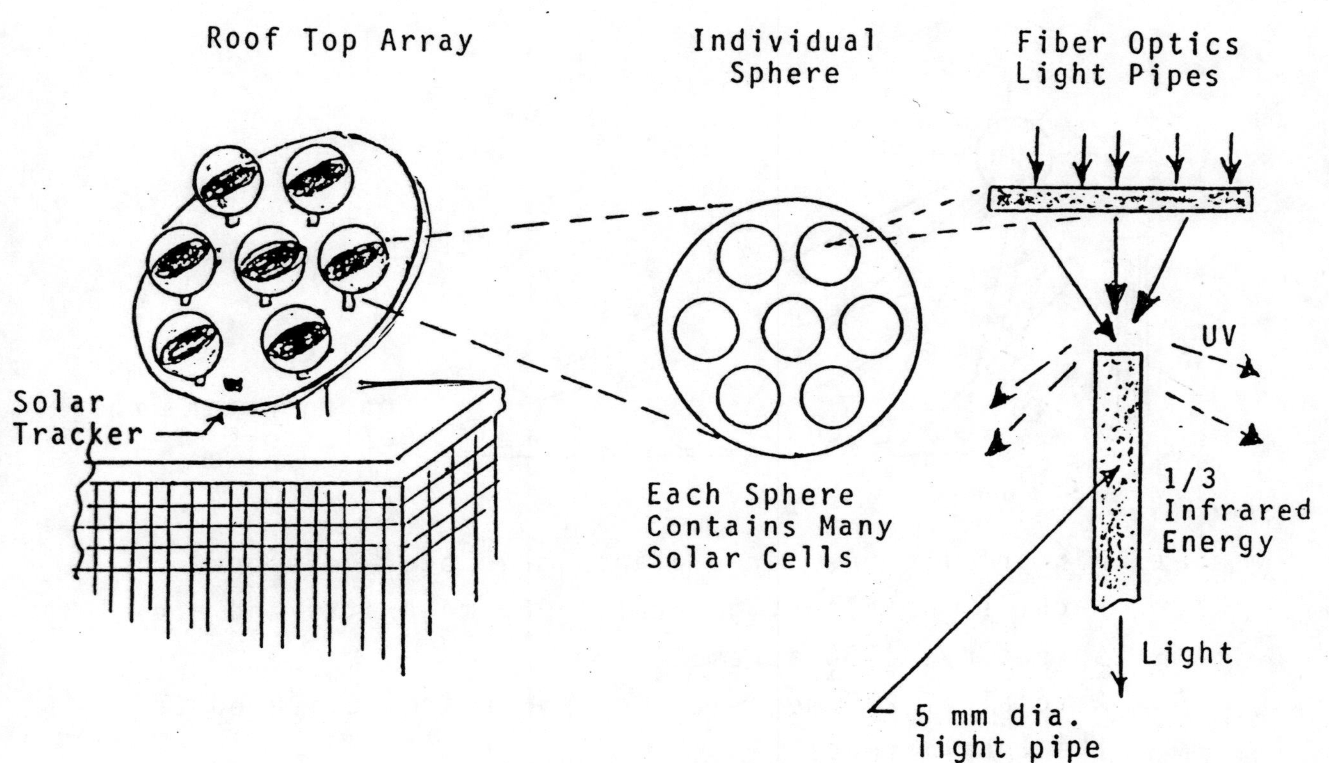

Specifics are the following:

- The system consists of multiple arrays, each system designed to track the sun.
- The system transmits all of the light into the basement of the building for gardens and medical purposes(vitamin D production in skin).
- Using Fresnel lens, 33% of the thermal energy and 100% of the UV energy is removed before the light reaches the millions of fiber optic light pipes.
- The system is expensive($120,000 per unit).

MEGAWATTS OF SOLAR POWER

Giant new solar plants are being built in the American South-west desert. They are capable of generating megawatts of solar power.

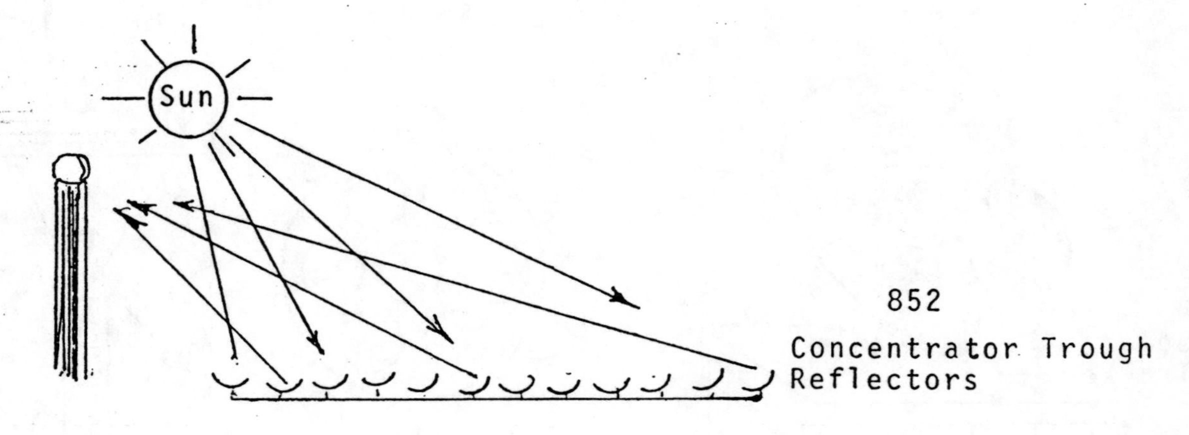

Specifics are:

- o Nine of the new large plants have been developed.
- o Each is capable of generating 80 megawatts and are spread over 380 acres.
- o A total of 275 megawatts of power are delivered to the local areas.
- o The plants are being built by Luz Engineering Corp.
- o Each of the new 80 megawatt plants shrinks the CO_2 emmissions by 325 million pounds annually when compared with oil-fired power plants.
- o Electricity is produced for 8-9¢/ kilowatt-hour and this is cheaper that nuclear. Also it approaches the average cost of electricity(6¢/kilowatt-hour).

Source: J. T. Johnson, Popular Sci., May 1990, pp. 82-85.

FUELS FROM THE SUN

NOW, IT IS POSSIBLE TO USE SOLAR ENERGY TO PRODUCE CHEMICAL FUELS USING A NUMBER OF METHODS:

- o PHOTOSYNTHESIS
- o HIGH TEMPERATURE CHEMISTRY
- o PHOTOVOLTAIC CONVERSION

As part of the high temperature conversion, one of three methods may be used:

Small solar furnaces - tens of kilowatts
Solar Concentrators - hundreds of kilowatts
Central receiver towers - thousands of kilowatts

An example of a solar dish is the one located at Stuttgart, Germany shown below in a sketch.

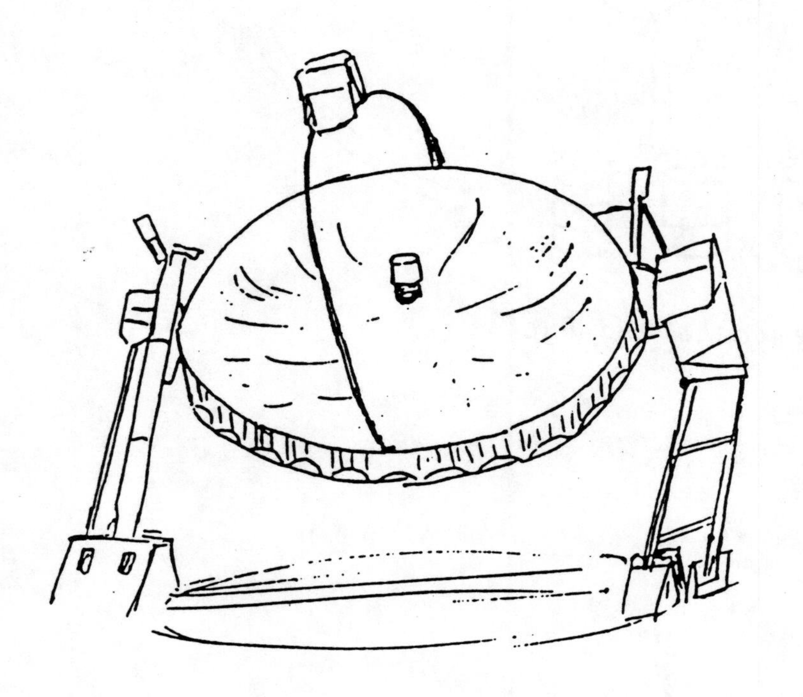

- o This dish intensifies solar radiation 10,000 times.
- o Power is 150,000 watts.

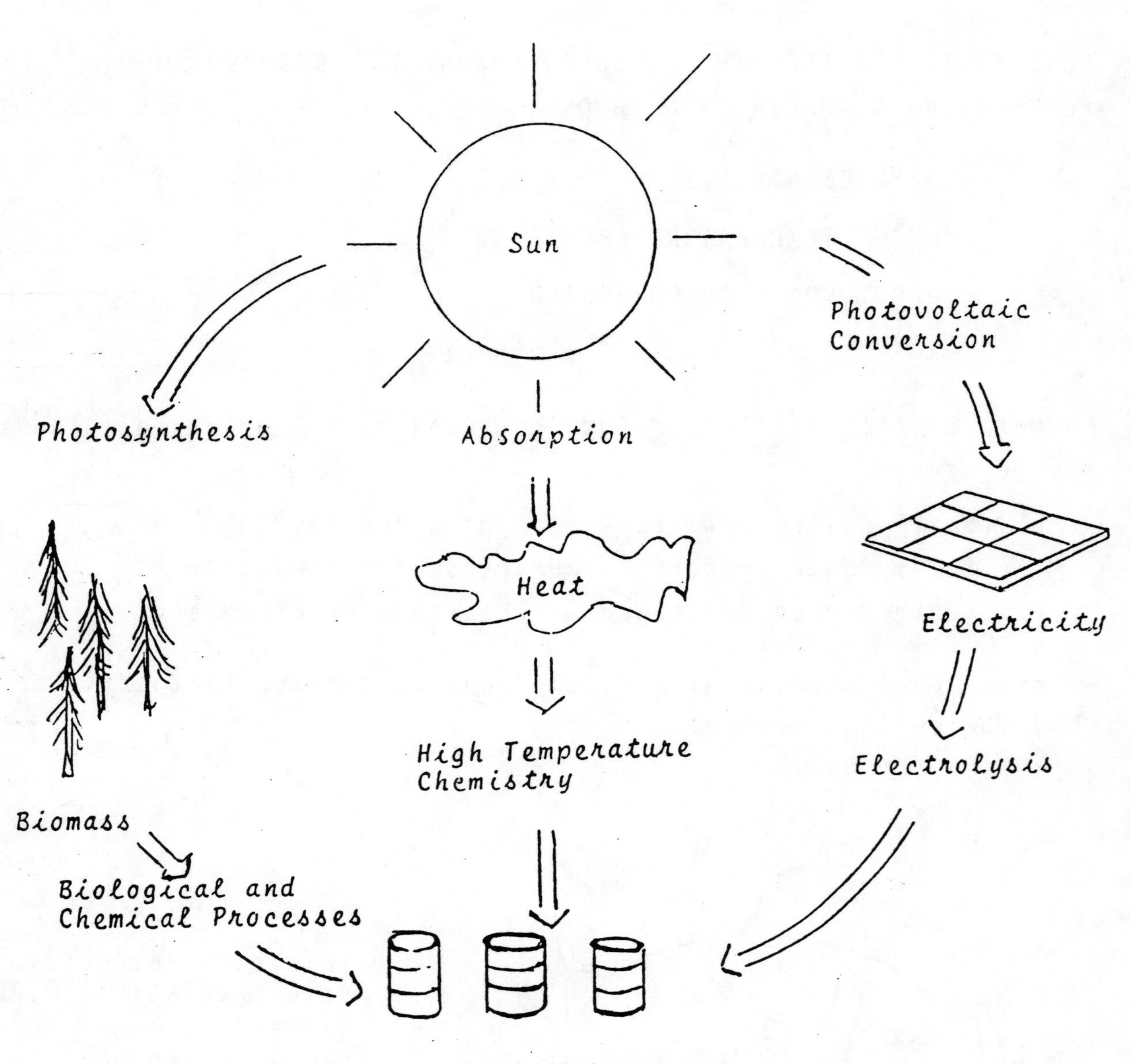
Sun
Photosynthesis
Absorption
Photovoltaic Conversion
Heat
Electricity
High Temperature Chemistry
Electrolysis
Biomass
Biological and Chemical Processes
Chemicals and Fuels

CHEMICAL FUELS

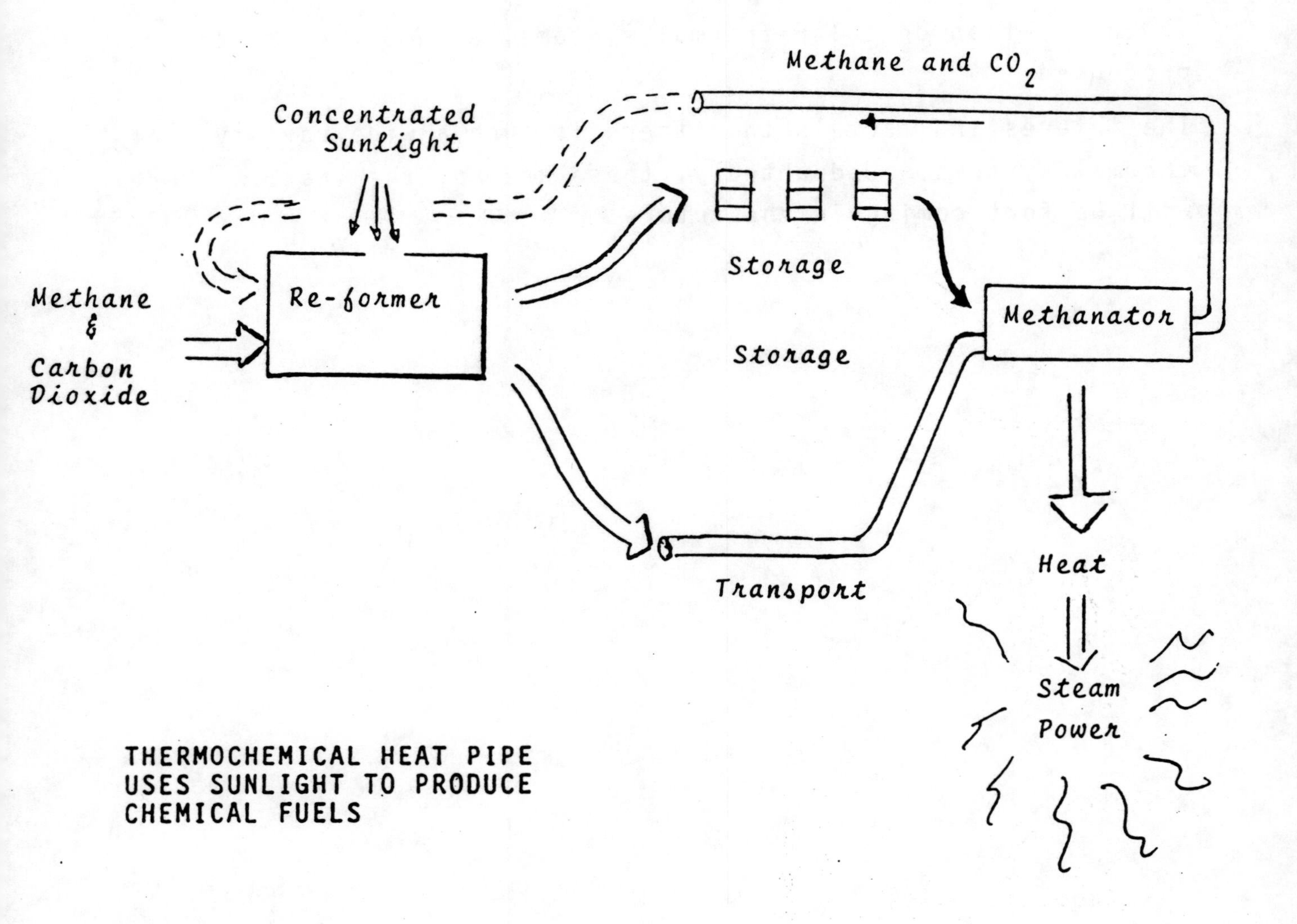

THERMOCHEMICAL HEAT PIPE
USES SUNLIGHT TO PRODUCE
CHEMICAL FUELS

Source: Dostrovsky, Israel, Chemical Fuels from the Sun, Scientific American, pp. 102-107, December 1991.

Summary

In the section on solar thermal systems, a number of types are presented.

The interesting fact is that there is such a wide variety of thermal systems. Undoubtedly, these are not the last. More will be forthcoming in the future.

Shrink-Fit Window Insulation Kit

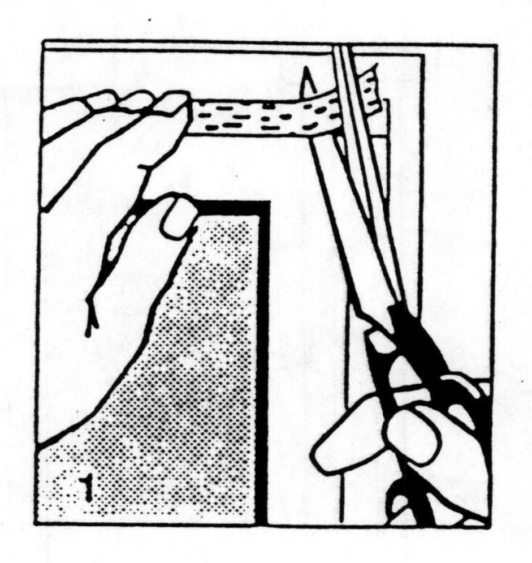

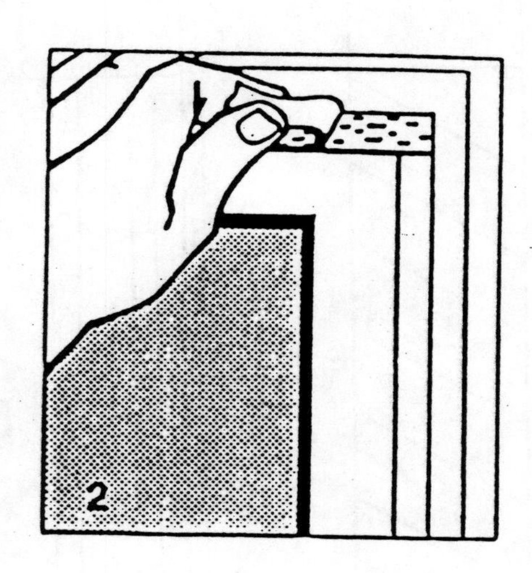

AVAILABILITY:

3-M 3-M Company, Box 33053
St. Paul, MN 55133-3055
Thermwell Products, 150 E. 7th St., Patterson, NJ 07524
Belleweather, BEDE INDUSTRIES, 8327 Clinton Rd., Cleveland, OH 44144

CONDUCTION LOSSES THROUGH
WINDOWS AND WALLS

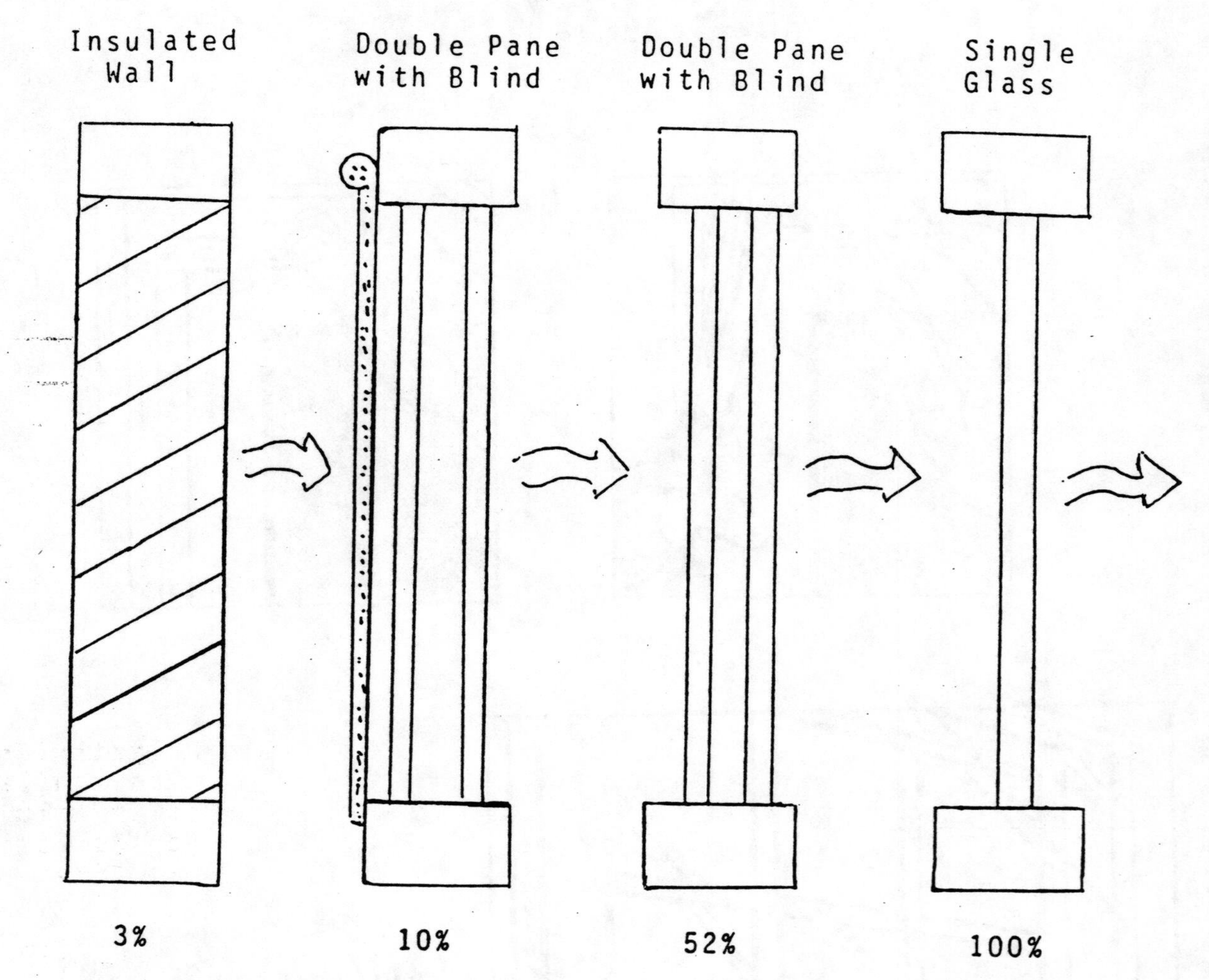

RELATIVE HEAT LOSSES THROUGH CONDUCTION

ADVANCES IN SOLAR DRYING TECHNIQUES

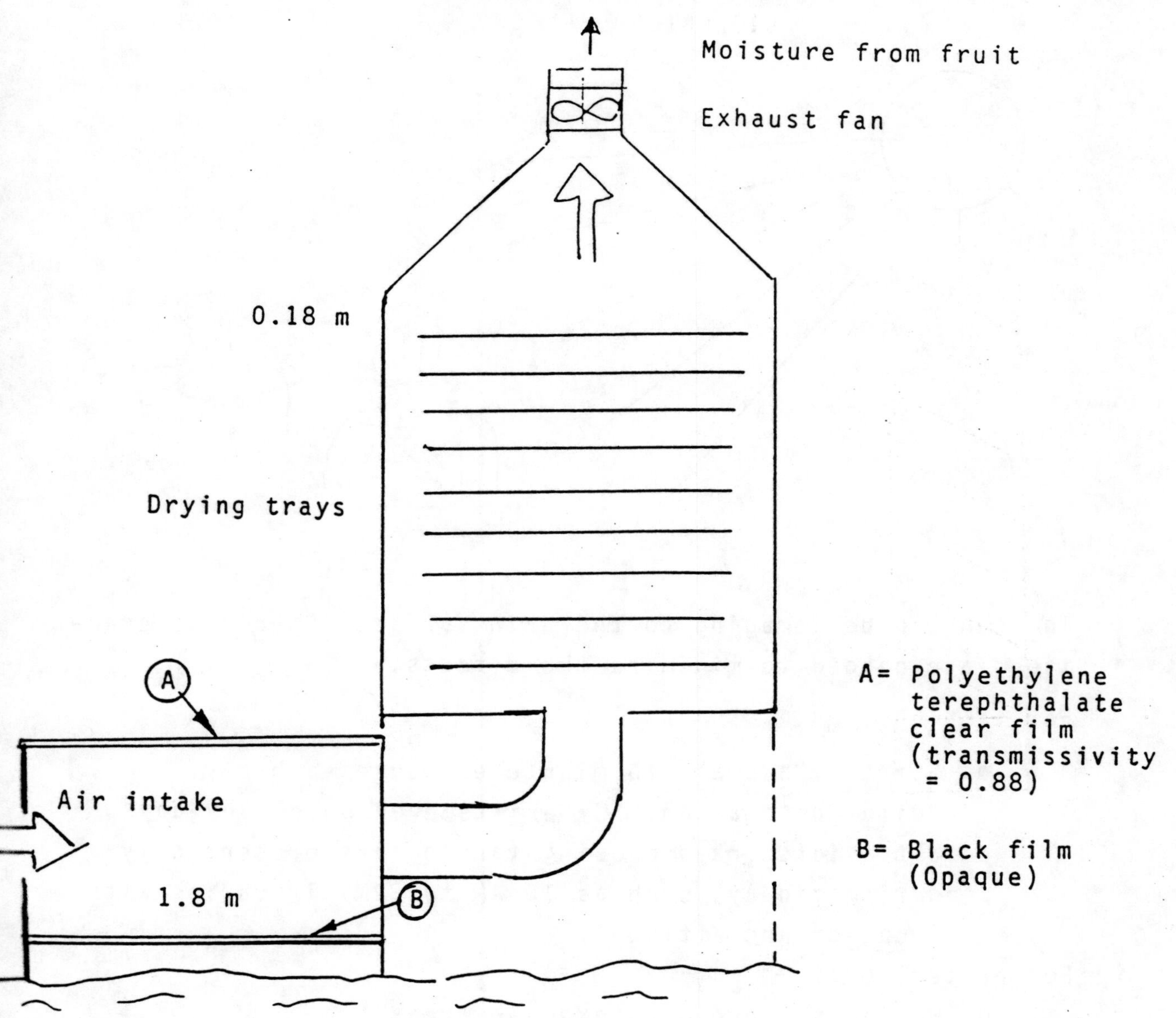

ADVANTAGES COMPARED TO ORDINARY DRYING

- o Increase in productivity is 2.5 times ordinary drying.
- o Output of finished product 28-30 percent ordinary.
- O Typical drying time for fruit(melons) is reduced from 11 days to 3 days.

REFERENCE: Applied Solar Energy, Vol.26,No. 2, 1990.

SOLAR RADIATION

BIOLOGICAL EFFECTS

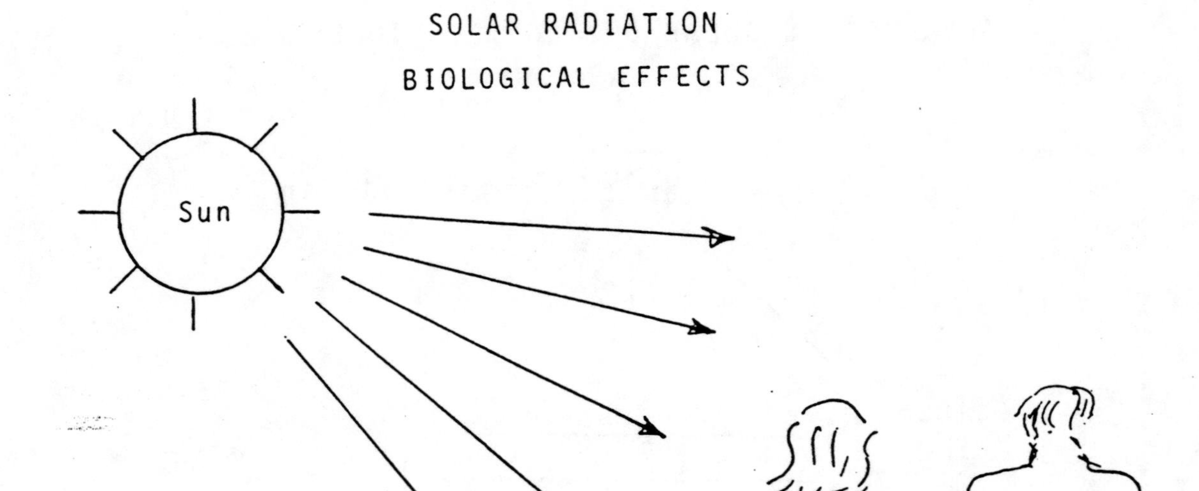

The sun can be damaging to the skin surface. Special precautions are needed to minimize the effects.

For tanning:

- o Start by a small 15 minute exposure each day. Build up to a tan. UV exists even on cloudy days.
- o Ultraviolet rays causing tanning are present only during midday, such as 10 AM to 3PM. It varies with location and altitude.

For protection:

- o Wear a sunscreen on exposed areas.
- o An opaque lotion is best against the ultraviolet.
- o Swim or sun bathe before 10 AM and after 3 PM.
- o Unprotected, solar effects could form on the epidermis later form on the lower dermis layer, and proceed on to cancer.

For more detailed information, see SOLAR ENERGY TECHNOLOGY-1991, by Systems Co.

SAFE SOLAR STRATEGY

It has been known for some time that sunlight, incident upon the human body is beneficial in small amounts. Examples include the elimination of rickets.

However, in larger amounts, it can be detrimental. Specifically, two types of effects exist:

- Growth of skin melanomas in later years due to childhood or early adult exposure. Damaged cells do not become visible until years after exposure.

- Excessive sun exposure for certain types of high-risk individuals:

  - Fair-skinned individuals.
  - People who are fair-skinned.
  - People who have many moles.
  - People who burn easily.
  - Individuals who have a family history of solar-induced skin disease.

People in the latter categories should wear hats, protective clothing, sunglasses, and use SPF-15 sunscreen.

SOLAR AUTO VENTILATION

An improved auto ventilation system has been invented by a company in the United States. It appears to satisfy a consumer need to "put the sun to work".

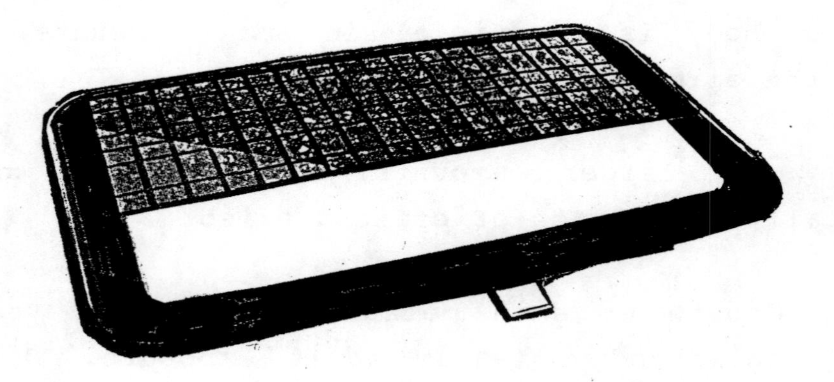

THE WEBASTO SUNROOF

Photovoltaic solar cells are used in the auto roof to keep the car cool.

Specifically:

- When the temperature made the car exceed 85° F, the system automatically activates.
- PV cells power two fans that pull in the cooler outside air and exhaust hotter air through vents in the roof.
- Should it rain, sensors shut the vents.
- Price: $1,300 (Dealer installed)

AVAILABILITY: Webasto Sunroofs
Maumee, Ohio U.S.A.

SOLAR ENERGY

Imitation Solar Energy

For parts of the United States where daily activities preclude exposure to the outdoor sunlight, artificial indoor sunlight has been invented. Over 35 million people are reported to suffer from lack of sunlight due to seasonal weather or working conditions.

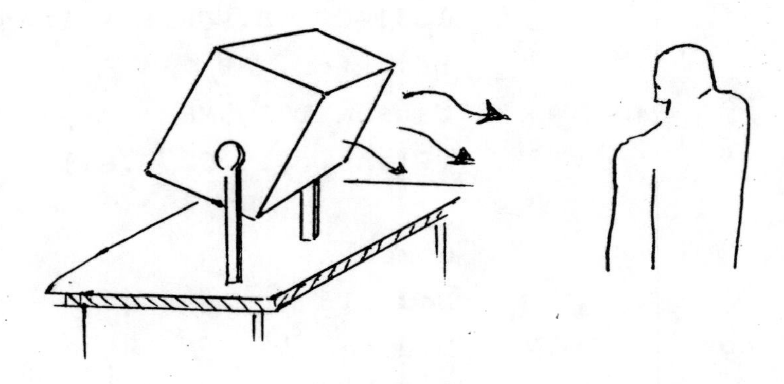

A small metal box emitting broad-spectrum fluorescent tubes has been invented.

Specifics:

- o "Sunray" is a 23 by 15.5 inch box with 4 lights.
- o Four light tubes per box.
- o Individuals report they feel better with 15 -30 minutes exposure per day.

Availability:

- o Neal Owens, President
  Sun Box Company
  Gaithersburg, Maryland
- o Price = $500.00

MISCELLANEOUS SOLAR PRODUCTS

There are many applications of solar power. Some are given below for the reader to pursue.

Type	Source
HYDROGEN AND OXYGEN GENERATOR	Praire Power 1331-194 S. Park Village Edmonton, AB Canada T6J 5M8 (Plans: $15.00 U.S.)
BACKYARD SOLAR ALCOHOL REFINERY (includes converting cars to alcohol)	HomeGen Box 815 Pleasanton, CA 94566 U.S.A. (Plans: $9.95)
VARIOUS SOLAR PRODUCTS	SUNIX CORP. P.O. Box 3889-PS San Rafael, CA 94912-3889 U.S.A. (Catalog: $5.00)

SOLAR-POWERED TOYS

A number of unique solar-powered toys are now available from a number of sources. These include:

- SUN-POWERED RADIOMETER ($19.95)
- SOLAR DIRIGIBLE ($4.95)
- SUN-POWERED CAR ($49.00)
- BUILD YOUR OWN SOLAR-POWERED LABORATORY ($39.95)

They are available from the following source:

EDMUND SCIENTIFIC CO.
Department 14D5
C915 Edscorp Building
Barrington, NJ 08007
U.S.A.

IV SOLAR THERMAL SYSTEMS

In this section, some of the solar thermal systems that have been considered for utilizing the energy of the sun are treated.

Many applications exist. Not all have been treated herein. However, the wide variety of thermal systems that have been studied and some that have actually been translated into hardware are shown.

Countries with ample solar energy, such as India and Arabia, are making even further progress in the development of practical solar thermal power systems.

The systems treated are highly dependent upon the success of materials that can withstand the effects of sunlight and other environments for extended duration. These aspects are treated in the following section.

Chapter 5

SOLAR MATERIALS

Crucial to the successful design of a successful solar energy system is the type and amount of heat resistant materials used. New materials and special design of combinations of materials are being designed continually.

Many materials and their properties do not change. These are listed along with the newer materials. Where possible, the sources and prices are given.

ENVIRONMENTAL EFFECTS ON MATERIALS

SOLAR REFLECTION

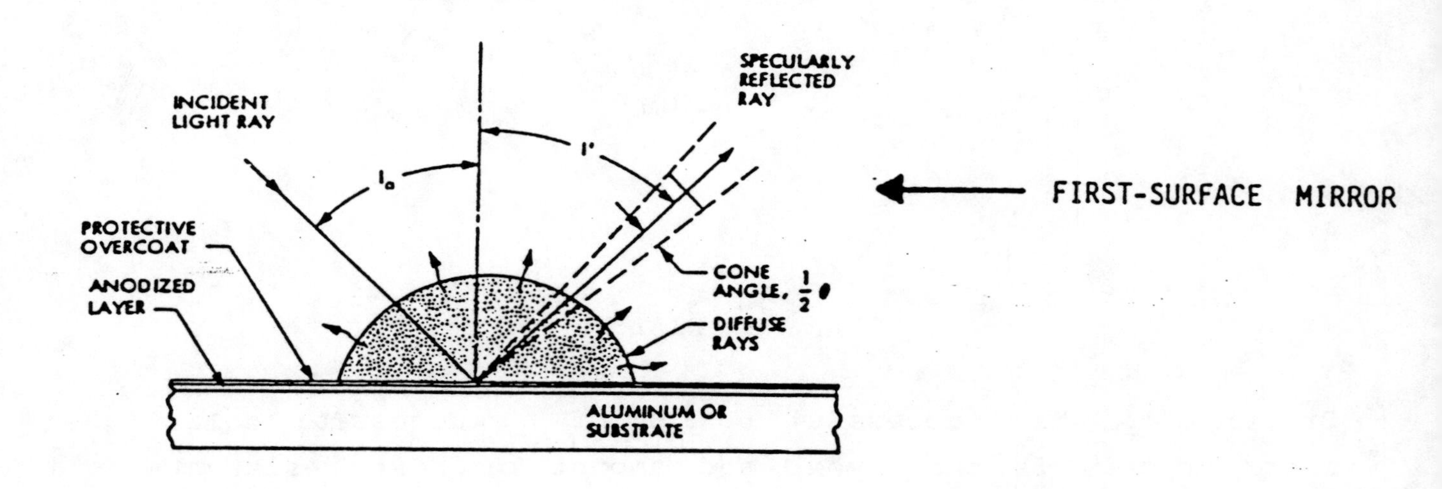

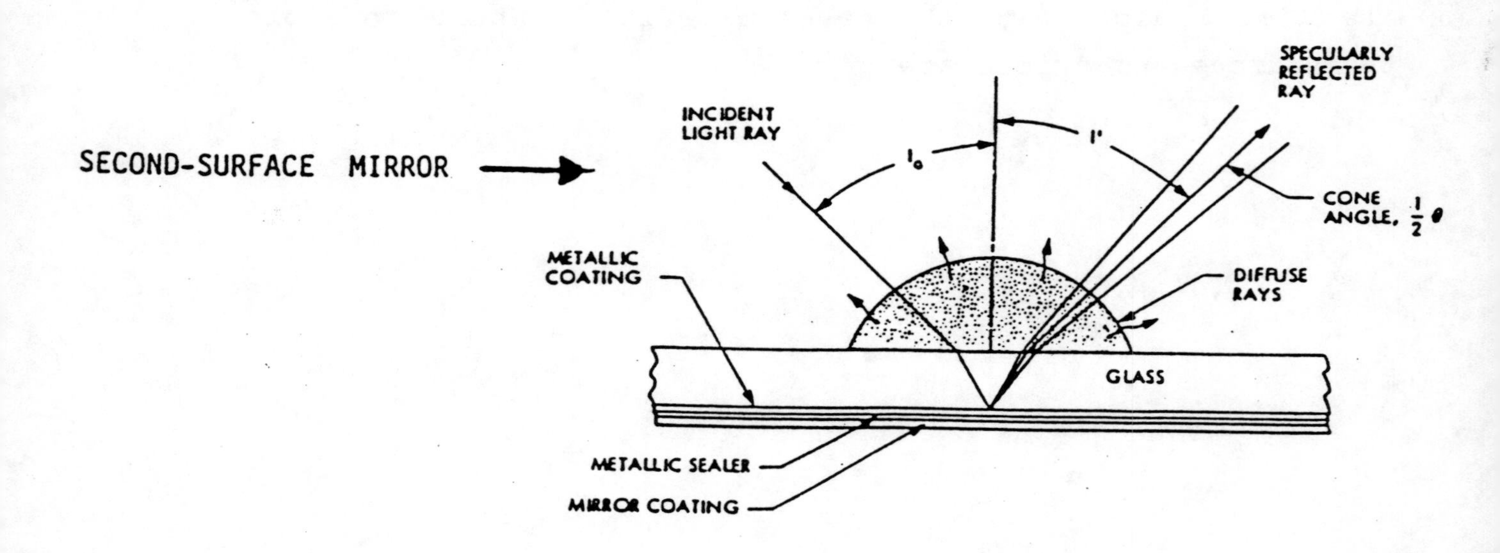

REFLECTOR SHEET MATERIALS

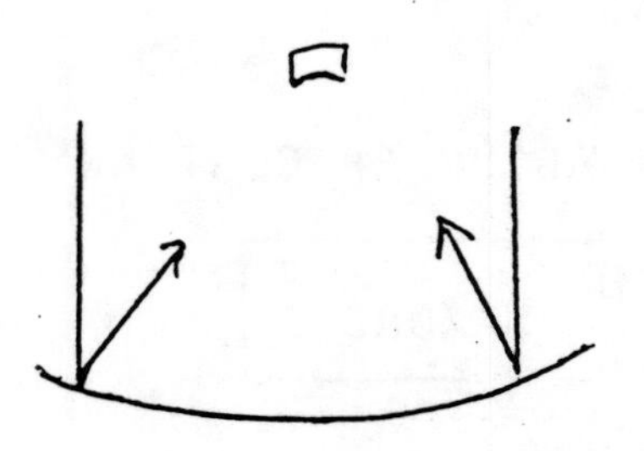

Various types

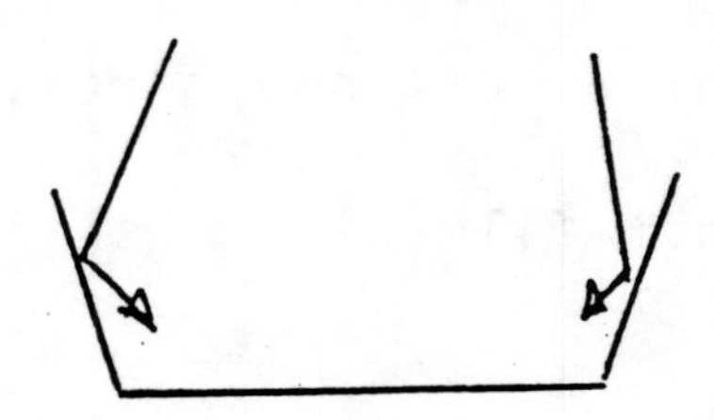

- Many types of geometrical reflectors are available
- Superior reflectance
- Anodized aluminum surfaces

SOURCE:

Kingston Industries Corp.
Kingston Way(S)
Sullivan International Airport
White Lake, NY 12786
914-583-5000

SOLAR FOILS

SPECIAL SOLAR ABSORBER COATING

MAXORB

- After extensive research, these special films were developed. Many types were tested.
- Available in a continuous roll.
- It is a black film with a silicone adhesive.

SOURCE:

NOVAMET
An Inco Company
681 Lawlins Road
Wyckoff, New Jersey 07481
201-891-7976

SOLAR REFLECTIVE MATERIALS - EFFECT OF AIR MASS

A word about first-surface solar reflective surfaces.

Solar engineers deal with three types of solar reflective surfaces, namely:

- Basic material surface reflectance. This is the scientifically measured reflectance of light perpendicular to a surface at a given wavelength. See the next page.
- Solar spectral reflectance. This is the total solar reflectance over all wavelengths impinging on a surface and being reflected perpendicularly.
- Milliradian reflectance. This is the total solar spectral reflectance impinging on a surface through a specified angle, and being reflected through another angle.

As a practical matter, the reflectance is frequently calculated or measured for a given air mass overhead. The larger the air mass, the greater the attenuation. See the sketch above.

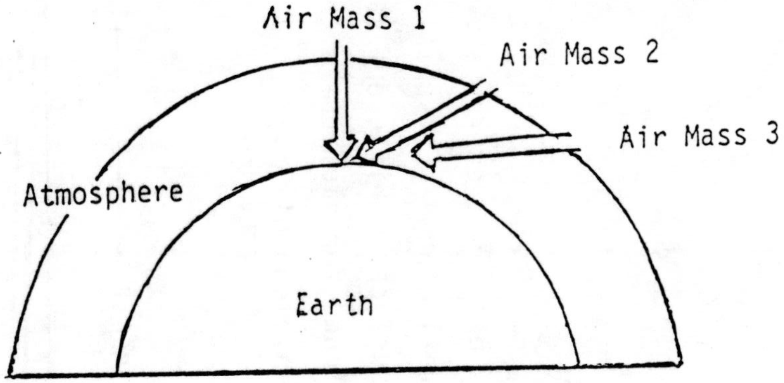

Air Mass 1
Air Mass 2
Air Mass 3
Atmosphere
Earth

SOLAR MATERIALS
- Spectral Reflectance

When using metallic reflectors, how do the solar rays reflect as a function of wavelength?

These basic curves show the reflectance of typical metal surfaces. Different values are found for aluminum at short wavelengths.

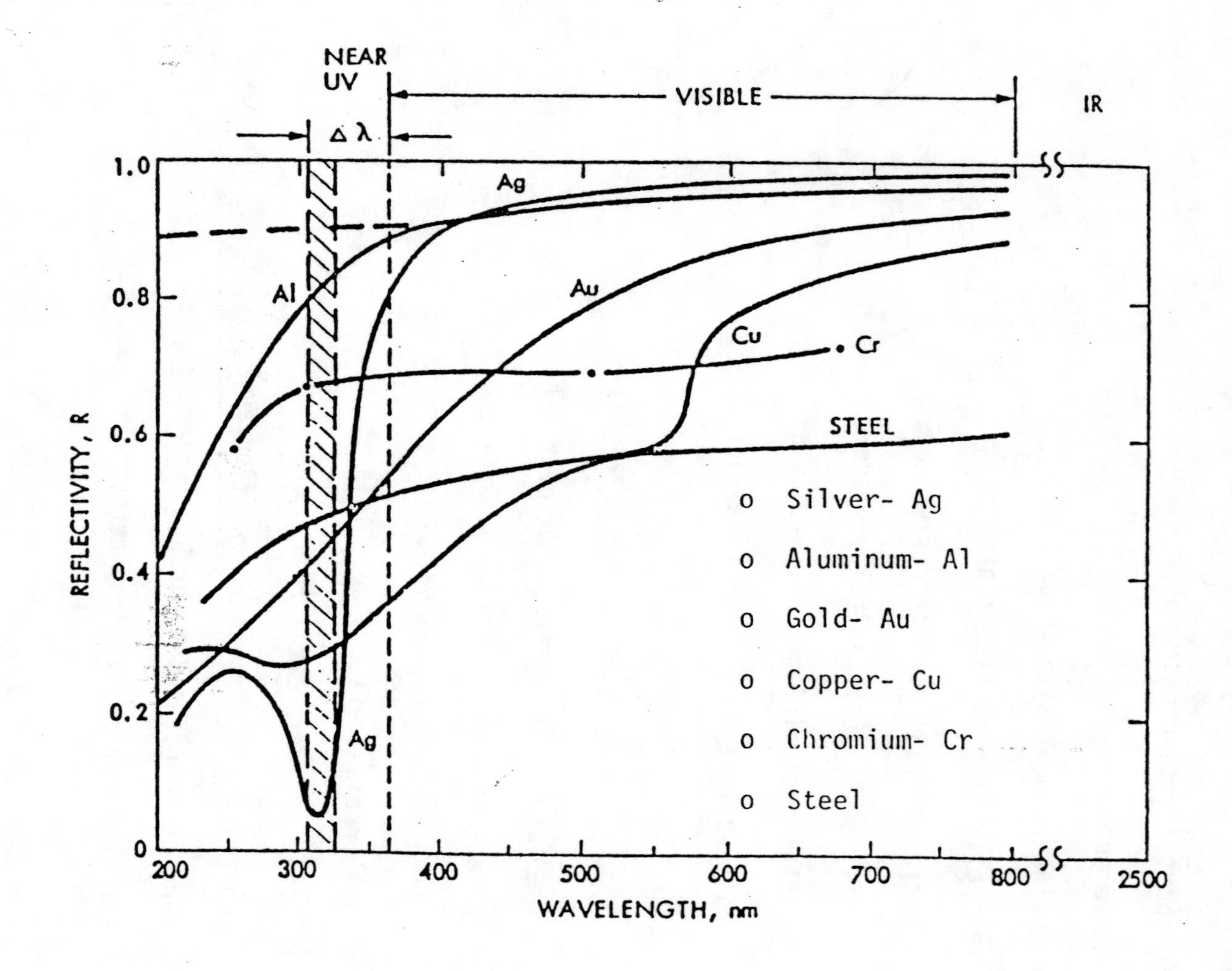

Solar Reflectance vs Wavelength for Various Metallic Surfaces

SOLAR MATERIALS

Relationship Between Theoretical and Measured Surfaces

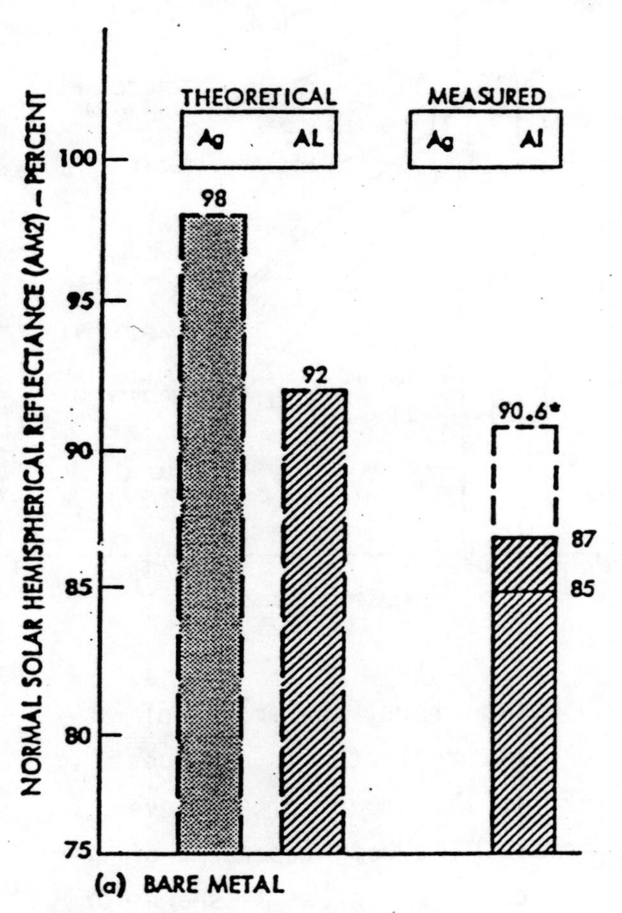

(a) BARE METAL

*DATA FROM KINGSTON INDUSTRIES, ANODIZED ALUMINUM

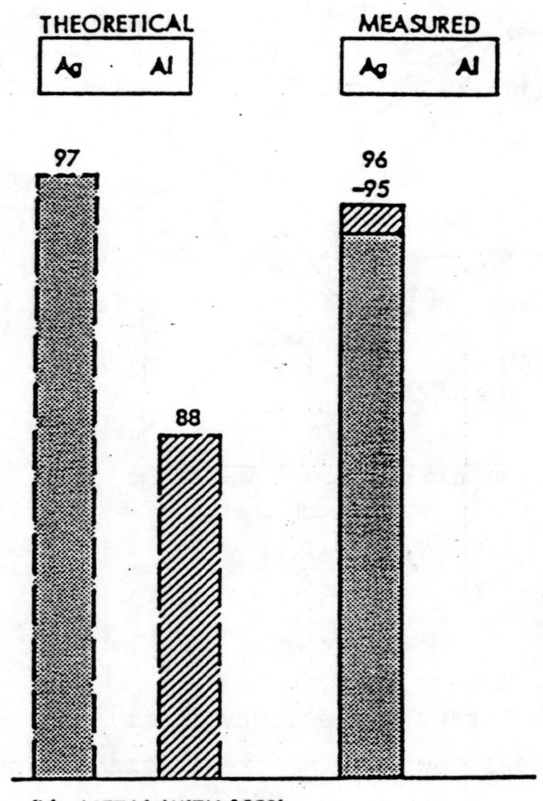

(b) METAL WITH 100% TRANSPARENT OVERCOAT

In this figure, the normal(perpendicular) radiation over the hemisphere is measured for very high refecting surfaces.

Air Mass Two(AM-2) is used because it is characteristic of the average value of the reflecting surface energy over the day.

The missing data have not been measured to date.

SOLAR MATERIALS

Glass Reflection and Transmission

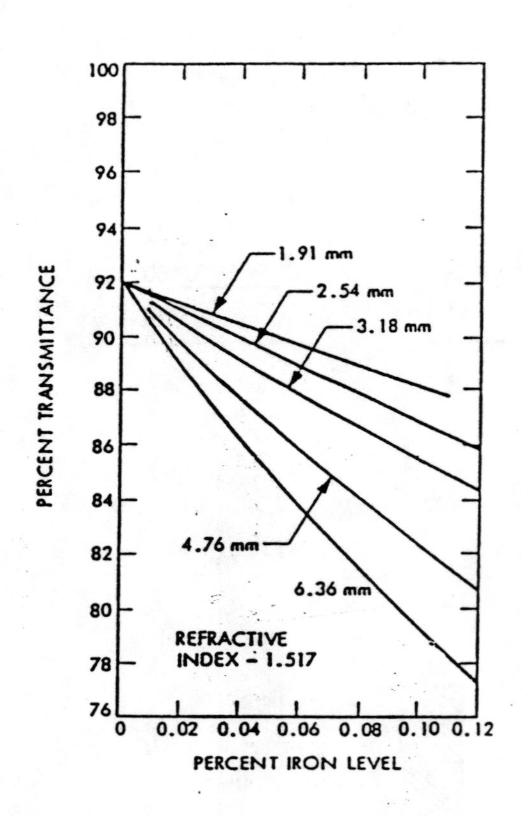

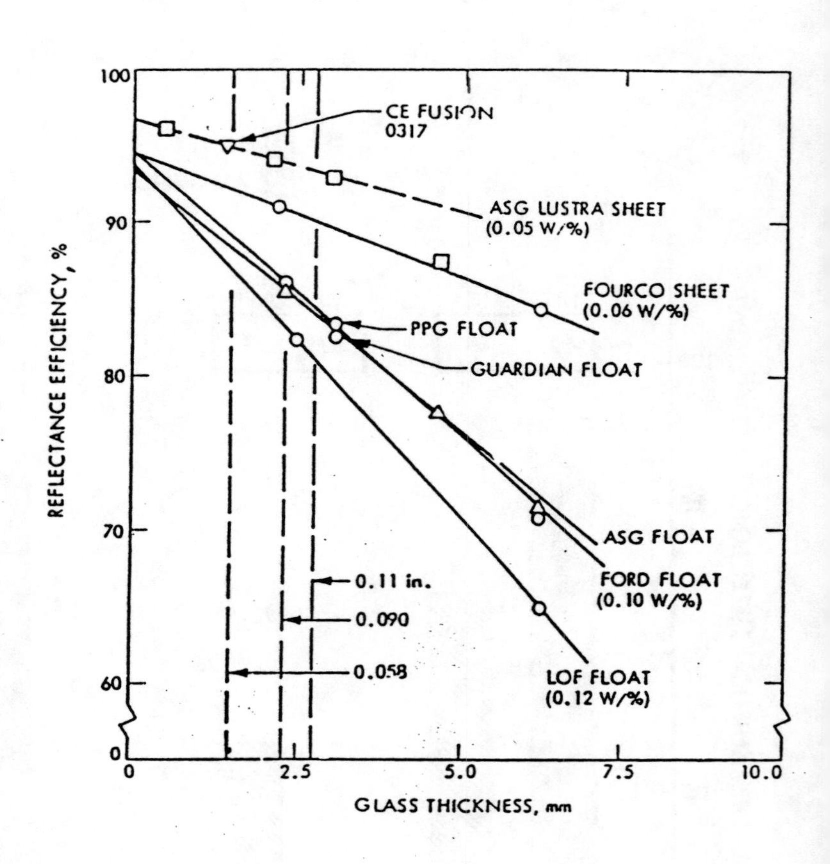

The transmission curve(left) shows how the the transmitted light falls off rapidly with the iron content of the glass. The greener the glass appears, the higher the absorption.

For mirrors, the amount of light reflected is dependent upon the amount of iron even more. The weight percent of iron in the glass is shown for various glass manufacturers for the 1982-1984 time period.

For mirrors, thin glass,such as 0.058 inches thick and low iron content are needed for very high reflectance.

SOLAR MATERIALS

Details of Construction of a Typical Mirror for Outdoor Exposure

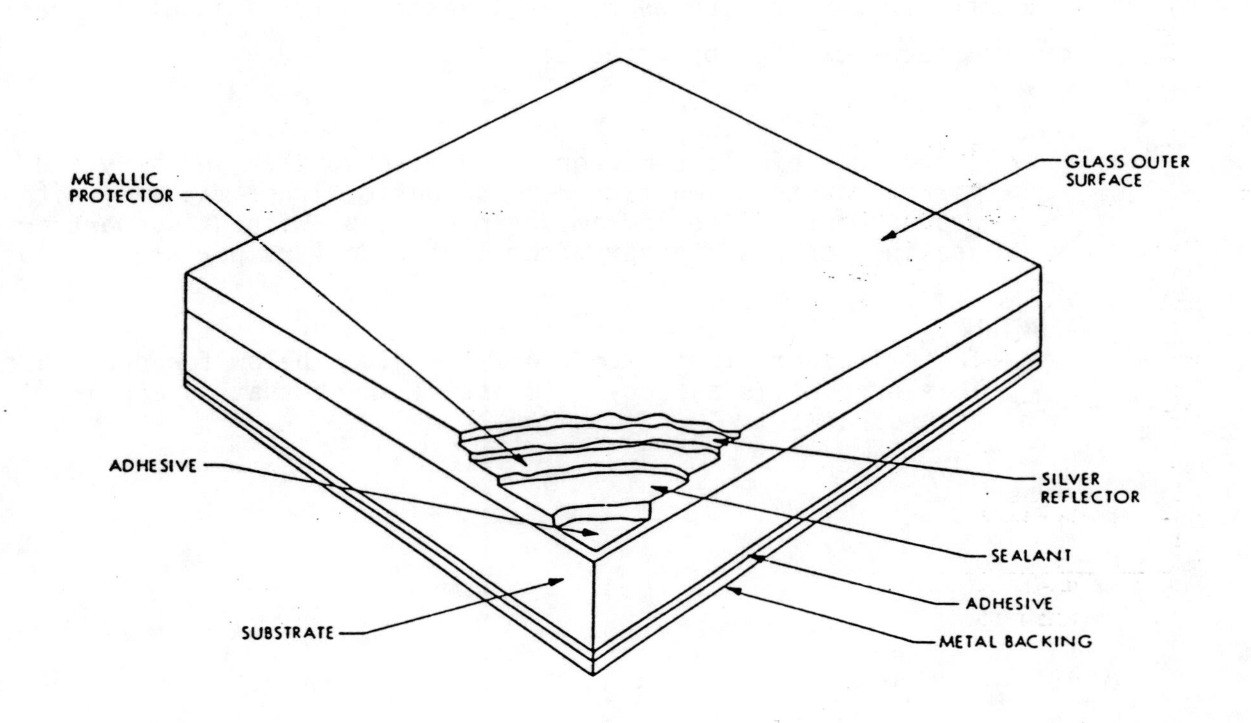

Many detailed mirror designs are available.

The important factors in design and construction are:

- Durability of the surface. A twenty year life is needed.
- Edge sealing to exclude moisture.
- Low weight.
- Low cost.
- Careful fabrication according to the specifications.
- Outer surface preparation to reduce dirt affinity.
- Surface compatibilty with cleaning agents to be used over the life of the mirrors.
- Mirror holders should be designed for easy replacement if damaged by hail or other phenomena.

PRICES OF REFECTIVE SURFACES

Prices are not so important for small home concentrators. For large concentrator surfaces, however, the investment for initial and replacement surfaces can be appreciable.

GLASS:

The prices of glass mirrors vary with complex industry and mirror construction factors. The basic mirror may vary with the procurement source and amount purchased. Without support or back sealing, prices may vary from 1 to 3 dollars per square foot.

ALUMINUM:

The prices of aluminum mirrors are given below for one source. This product is an acrylic(plastic) sheet that is applied to a substrate.

Area Purchased		Price(1980 $) per Unit Area	
m^2	ft^2	$\$/m^2$	$\$/ft^2$
0.93	10	18	1.67
9.3	10^2	18	1.67
93.0	10^3	15	1.39
930.0	10^4	13.23	1.23
9,300.0	10^5	13.23	1.23
93,000.0	10^6	13.23	1.23

ENVIRONMENTAL RESPONSES OF MATERIALS

HAIL

If a solar concentrator is designed for operation in areas where frequent hail is expected, sensitive glass surfaces can undergo damage. The amount of damage is related to the response of the substrate as well as the impacted material. The force of the hail may make permanent indendations in the substrate as well as the reflective surface. Reflection properties of the mirror may degrade beyond specifications.

In brittle substances, such as glass, hail can cause significant degradation.

HAIL IMPACT DAMAGE

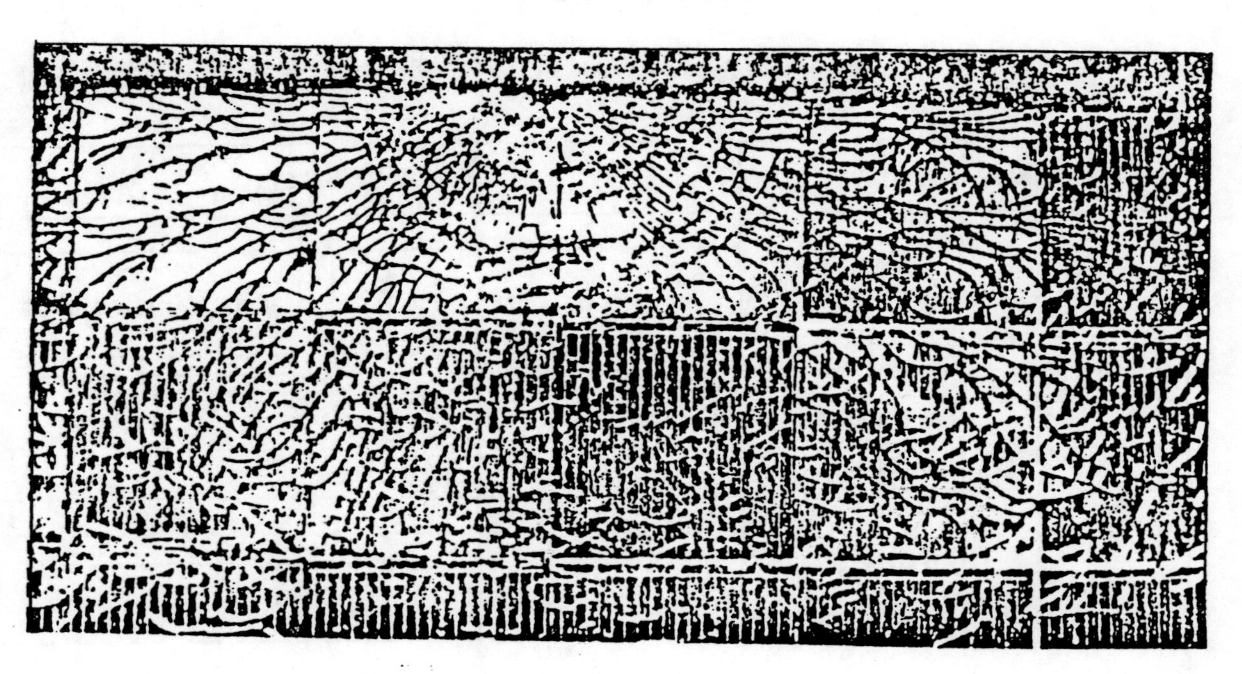

Reference: Progress Report 13, DOE/JPL-1012-29, CIT, Pasadena, CA, August 1979

ENVIRONMENTAL RESPONSES OF MATERIALS

HAIL

If a solar concentrator is designed for operation in areas where frequent hail is expected, sensitive glass surfaces can undergo damage. The amount of damage is related to the response of the substrate

Both acrylic plastic sheet and heat tempered glass are superior in withstanding hail impact, compared to soda-lime annealed window glass.

The shaded region denotes where frequent failures will occur.

Adapted from D. Moore, Photovoltaic Solar Panel Resistance to Simulated Hail, JPL 5101-62, California Inst. of Tech., October 15, 1978.

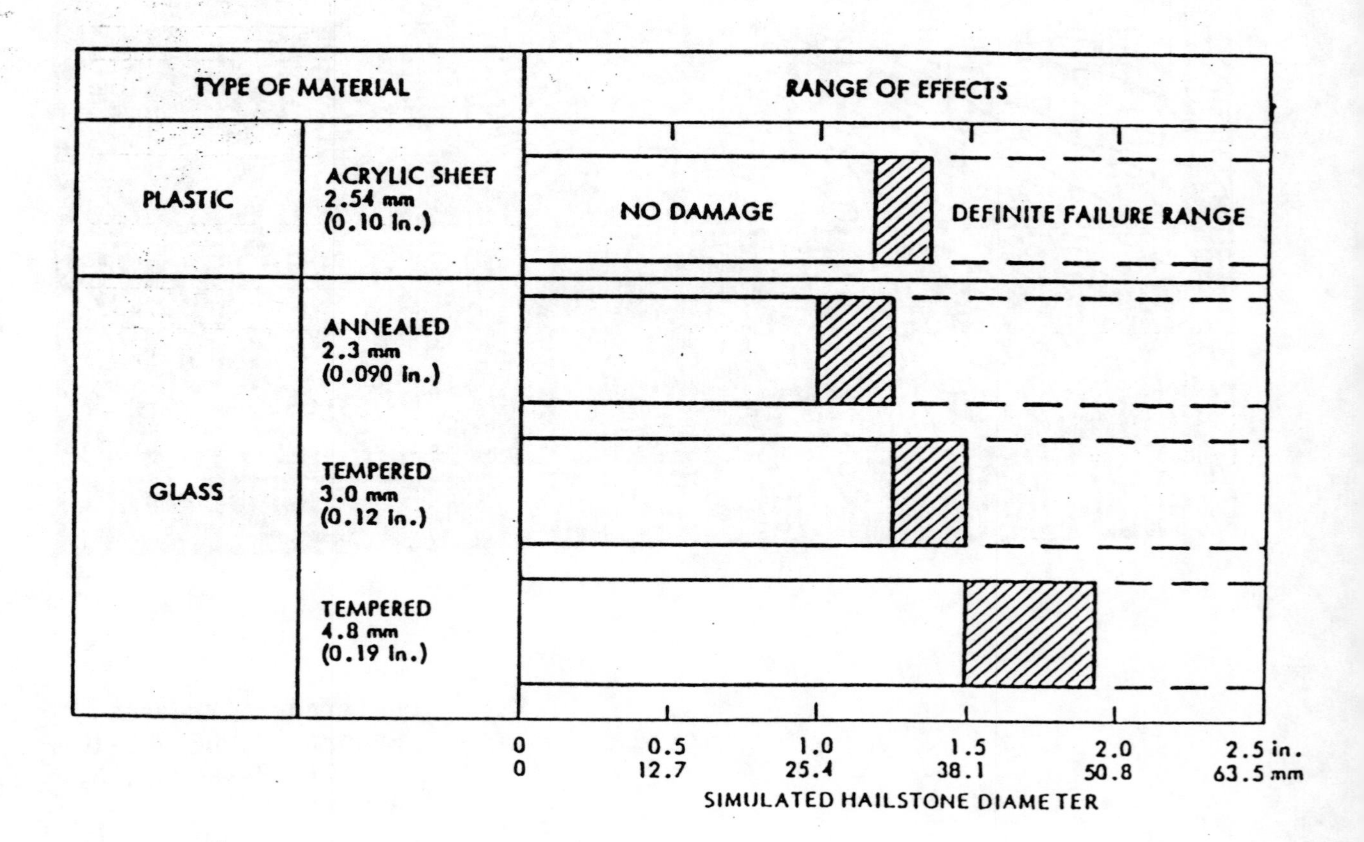

SOLAR ELECTRIC POWER

Summary of the Status of the Technology Base for Large Parabolic (Two-dimensional) Disks- Reflective Surfaces.

Type of Reflector	Criteria									
	Temperature					Special Transportation Requirements		Optical Performance Screening Tests		Degradation and Cleaning
	Process Bending		Operating Levels	Gradient	Cycling					
	Hot	Cold				Handling	Shipping	Lab	Outdoor	
Silvered Glass Mirrors										
• Glass	●	○	○	⊕	○	⊕	⊕	⊕	⊕	⊕
Metallization										
• Silver			●	⊕	●	⊕	⊕	⊕	●	
• Copper			●	⊕	●	⊕	⊕	⊕	●	
• Paint			●		●	⊕	⊕	⊕	●	
Aluminum										
• Anodized Layer		⊕	⊕	○	●	○	○	⊕	●	●
• Bulk Aluminum		⊕								
Metal Film										
• Polymer		⊕	⊕	⊕	●	○	○	⊕	●	●
• Film		⊕	⊕	⊕	●	○	○	⊕	●	

○ Technology base exists
⊕ Partial technology base exists
● Technology is not well known

PROTECTION AGAINST CORROSION

All outdoor systems, solar or non-solar, need to be protected against premature life-shortening. Proper design and installation are all important factors in prevention.

Examples of important aspects to be considered.

- Do not mix metal pipes of different types. Make sure that there are no galvanic couples to produce corrosion.
- Avoid stagnation temperatures in the coolant flows.
- Soldering flux should be flushed from the installed system shortly after installation(within a few days).
- In cold climates, use special pumps to avoid freeze conditions.
- Exercise care in switching the types of fluids in the system to avoid incompatibility.
- Control the pH of the system and reduce calcium carbonate scaling.

Other efforts to reduce corrosion should be taken both during installation and periodic maintenance. For further information see the article by Avery and Krall.

REFERENCE: John Avery and John Krall, Extend Lifetimes of Systems by Reducing Corrosion, Solar Engineering & Contracting, Vol. 1,No.2, February 1982.

S O L A R   R E F L E C T O R S

-Glass, Second Surfaces

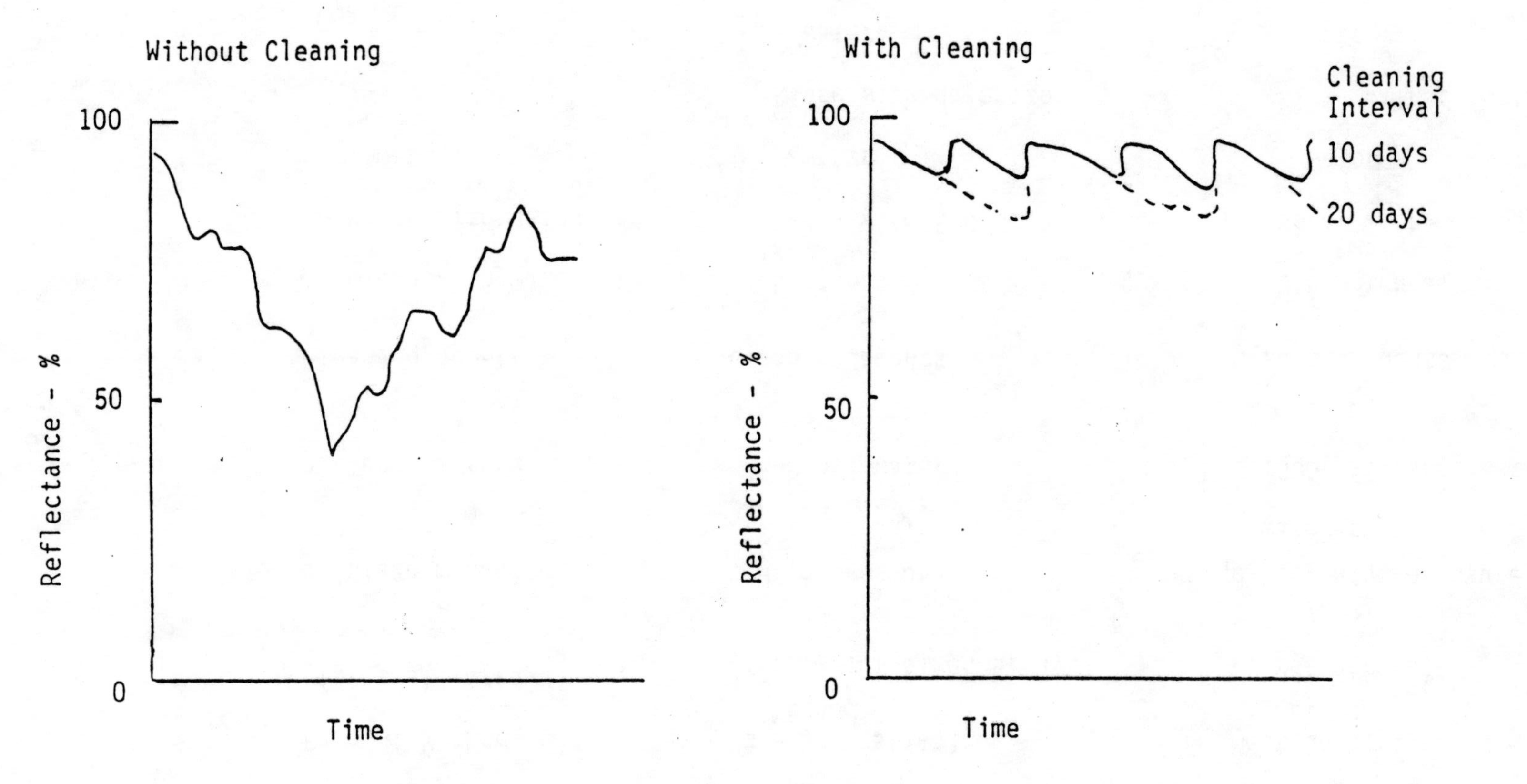

Typical Profiles Measured over a One Year Time Period in Pasadena,California

CLEANING METHODS

Applicable to both PVs and reflectors.

Many different methods are currently in use:

Easiest Method (for small arrays)	More Sophisticated Method (for medium-sized arrays)	Most Sophisticated Method (large arrays)
o Clean manually	o High pressure water spray	o Very high pressure water spray
o Tap pressure	o 500 psi water	o 1000 psi range water
Cleaning agents	Cleaning agents	Cleaning agents
Clean with water	Use soap	Use methonal
Rinse with deionized water.	Spray with 500 psi water	Spray with ~1000 psi water
	Rinse with deionized water.	Rinse with deionized water.

FUTURE SOLAR WINDOWS

ELECTRICAL

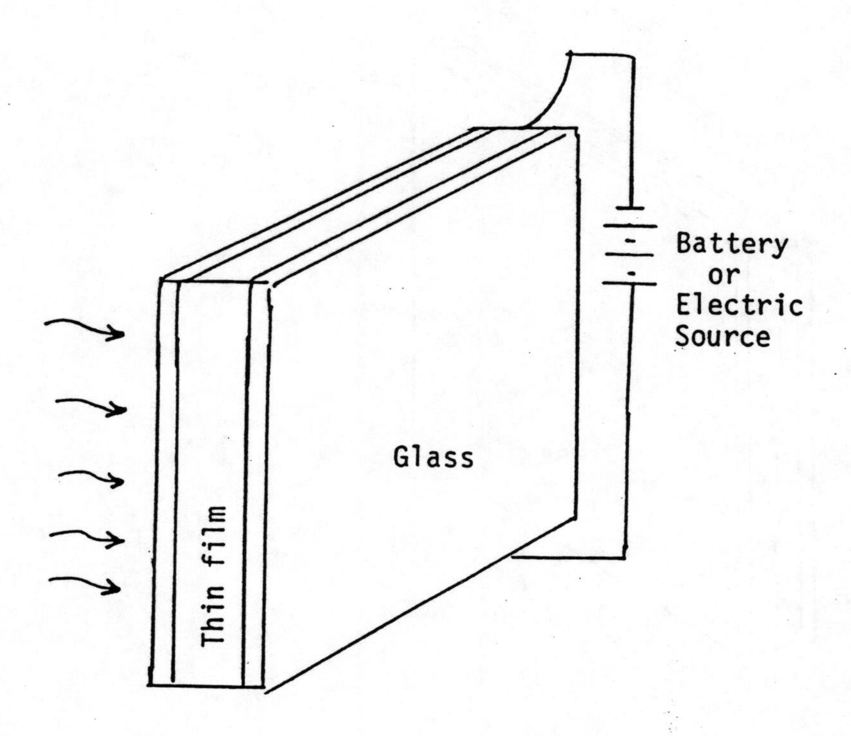

Opaque film changes color after voltage is applied. Turns black at night to trap heat.

SOURCE: Plastics that Conduct Electricity, Scientific American, Volume 258, No. 2, page 110, February 1988

FUTURE SOLAR WINDOWS (Cont.)

CHEMICAL

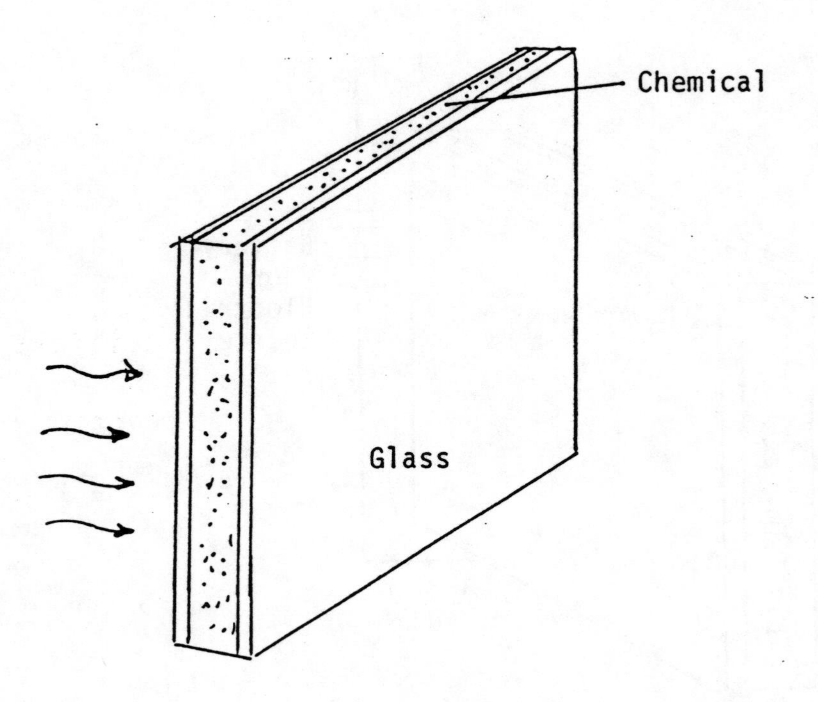

Automatically turns opaque at night for heat control.

Prototypes already developed.

SOLAR GLAZING

PANELS HELP HEAT HOMES

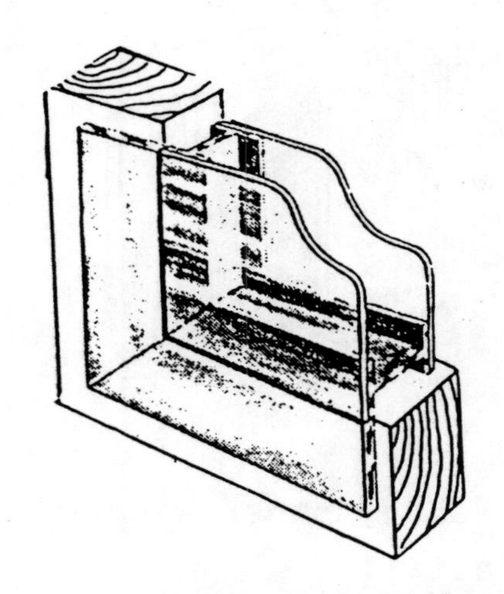

- Useful for greenhouses or solar house additions
- Offer maximum light and energy transmission

Source:

Vegetable Factory, Inc.
P.O. Box 2235, Dept SA-83P
New York, NY 10163

SELECTIVE SURFACE

WINDOW

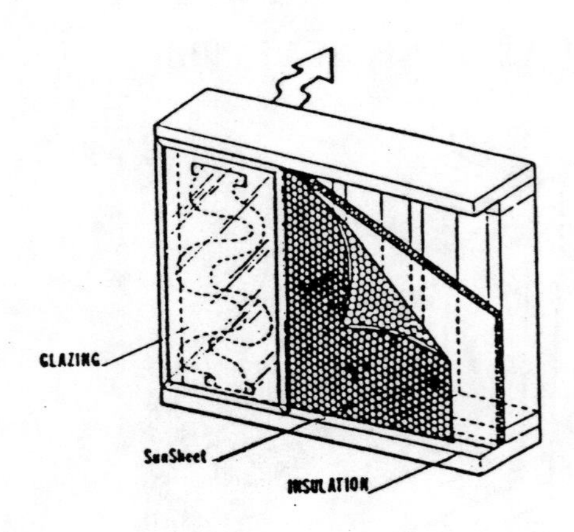

- O Looks like any other window
- O Easily installed
- o System payback in 4-7 years.

SUNSHEET™ comes in 48 inch widths

Available in rolls

SOURCE:

BERRY SOLAR PRODUCTS
A division of the Berry Group
2850 Woodbridge Avenue
Edison, New Jersey
201-549-0700

MATERIALS

Absorptance to Emittance Ratios

Substance	α	ϵ	α/ϵ
White plaster	0.07	0.91	0.08
Snow, fine particles, fresh	0.13	0.82	0.16
White paint on aluminum	0.20	0.91	0.22
Whitewash on galvanized iron	0.22	0.90	0.24
White paper	0.25-0.28	0.95	0.26-0.29
White enamel on iron	0.25-0.45	0.90	0.28-0.50
Ice, with sparse snow cover	0.31	0.96-0.97	0.32
Snow, ice granules	0.33	0.89	0.37
Aluminum oil base paint	0.45	0.90	0.50
Asbestos felt	0.25	0.50	0.50
White powdered sand	0.45	0.84	0.54
Green oil base paint	0.50	0.90	0.56
Bricks, red	0.55	0.92	0.60
Asbestos cement board, white	0.59	0.96	0.61
Marble, polished	0.5-0.6	0.90	0.61
Rough concrete	0.60	0.97	0.62
Concrete	0.60	0.88	0.68
Grass, wet	0.67	0.98	0.68
Grass, dry	0.67-0.69	0.90	0.76
Vegetable fields and shrubs, wilted	0.70	0.90	0.78
Oak leaves	0.71-0.78	0.91-0.95	0.78-0.82
Grey paint	0.75	0.95	0.79
Desert surface	0.75	0.90	0.83
Common vegetable fields and shrubs	0.72-0.76	0.90	0.82
Red oil base paint	0.74	0.90	0.82
Asbestos, slate	0.81	0.96	0.84
Ground, dry plowed	0.75-0.80	0.70-0.96	0.83-0.89
Linoleum, red-brown	0.84	0.92	0.91
Dry sand	0.82	0.90	0.91
Green roll roofing	0.88	0.91-0.97	0.93
Slate, dark grey	0.89	–	–
Bare moist ground	0.90	0.95	0.95
Wet sand	0.91	0.95	0.96
Water	0.94	0.95-0.96	0.98
Black tar paper	0.93	0.93	1.0
Black gloss paint	0.90	0.90	1.0
Small hole in large box, furnace or enclosure	0.99	0.99	1.0
"Hohlraum," theoretically perfect black body	1.00	1.0	1.0

Chapter 6

SOLAR RIGHTS

Solar rights are important if you are to go to the expense of installing a permanent solar system on your property. Others who may block the sun from your system would defeat your goals. The right to solar access has been under scrutiny in the legal world for some time. It behooves the designer or installer of a system to research the solar rights that he will minimize the probability of being blocked.

Many new developments have occurred, are occurring, in the field of solar rights. They vary with country and state.

Access to sunlight for running of solar energy devises is a most important aspect of solar energy economics. Who will install solar energy systems if they are to be blocked by trees, high rise buildings or other obstructions?

Consequently, there is great interest in the solar rights of access. As of today, the courts have only begun to review the legal ramifications of solar rights.

States that stood behind the solar rights for the individual by passing legislation are the following:

abbreviation	state	abbreviation	state
CO	Colorado	MD	Maryland
CT	Connecticut	MN	Minnesota
FL	Florida	NJ	New Jersey
GA	Georgia	NM	New Mexico
ID	Idaho	ND	North Dakota
KS	Kansas	OR	Oregon
		VA	Virginia

Information on the guidelines and laws concerning solar rights may be obtained from the Energy Department of your state or local government. They range from simple to complex.

For further information concerning legal decisions and court cases, see the References and Bibliography.

For more detail on the status of solar rights, see the References by Becker and Zillman below.

- R. Becker, Jr., Common Law Rights....., Jour. of Contemporary Law, Vol. 3, pg. 19, 1976.
- R. Zillman and R. Deeny, Legal Aspects of Solar Energy Development, Arizona State Law Journal, pg. 25, 1976.

SUMMARY

The status of the solar rights can be summarized as follows:

- There is a legal precedence "doctrine of ancient lights" that in English Law prescribes that if light, air and view have been enjoyed from a window for some 20 years, it becomes a right.
- Courts have ruled for and against this right over the last 150 years in various cases. Frequently, owners were compensated for loss of sunlight, such as from overhead railroad lines.
- As mentioned previously, some thirteen states have enacted some form of solar right(or solar access) guarantees. For information concerning the action of these states and greater detail on the entire rights question, see the treatment by Rapp.

VI SOLAR RIGHTS

An analogy has been drawn between water rights and solar access rights. However, it is noted that the former varies with the different states.

Considerable more needs to be done to clarify the rights of the solar energy user.

In the meantime, the laws cannot grant solar property owners complete freedom of solar access. It pays to follow closely the tested and established practices of common and civil law.

Chapter 7

CONVERSIONS AND DEFINITIONS

Because solar energy units are unfamiliar to most, important conversion factors are given in Table 7-1.

Likewise, the terms used in solar energy technology may be unfamiliar. A list of the more important terms are given in Table 7-2.

There are many engineering definitions used because of the many technologies involved. For greater detail, see Mark's Mechanical Engineer's Handbook, the books by Ted Lucas, or Bruce Anderson's The New Solar Handbook. All are listed in the References and Bibliography.

HOME OR BUSINESS SAVINGS CALCULATIONS

HOW DO YOU DECIDE IF A SOLAR ENERGY SYSTEM IS A GOOD INVESTMENT?

IS IT THE BEST WAY TO SPEND YOUR FUNDS?

There are three methods for evaluating this problem.

- SIMPLE PAYBACK METHOD
  This is the easiest to calculate.

$$\frac{\text{Total cost}}{\text{Annual return}} = \text{Number of years to pay back the investment}$$

- NET PRESENT VALUE METHOD
  This method takes the value of money into account.
  A dollar earned today is worth more than one earned five years from now.
  A simple calculation.
  Discounted rates of return information is available in most financial books.

- BENEFIT-TO-COST RATIO METHOD
  This method includes taking the value of money into account also.
  The rate of return on each dollar invested in calculated.
  A benefit-to-cost ratio greater than 1.0 is a positive return on investment.
  Using this method, the various systems can be compared on a like basis.

CONVERSIONS

It is frequently convenient to use solar energy conversion factors(c) from English units to the Standard International(SI) units. These are given in the following table.

Units	SI	c*
pound	kilogram	0.4536
foot	meter	0.3048
^oF	^oK	$(5/9)(T(^oF) - 32) + 273.16$
lb/ft^3	kg/m^3	16.02
ft-lb/sec	N(newton)	4.448
lb/in^3	Pa(pascal)	6895
BTU	J(joule)	1055
BTU/hr	W(watt)	0.2931
BTU/lb^oF	$J/kg\text{-}^oK$	4187
BTU/lb	J/kg	2326
$BTU/hr\text{-}ft^2$	W/m^2	3.155
ft^2	m^2	0.0929
gal	m^3	0.00378
^oF	^oK	0.555

*Note: c is the conversion factor to go from column 1 above to column 2.

BLANK PAGE

Table 7-2

DEFINITIONS

Absorber	Component of a solar collector, usually metallic, whose function is to collect and retain as much of the radiation from the Sun as possible.
Absorptance	The ratio of solar energy absorbed by a surface to the solar energy striking it.
Active system	A solar heating or cooling system that requires external mechanical power to move the collected heat.
Air Mass	The length of the path of the solar radiation through the atmosphere by the direct radiation as a multiple of the path length with the Sun at the zenith (vertically overhead).
Ambient	The surrounding atmosphere. Thus, ambient temperature means the temperature of the atmosphere at a particular location.
ASHRAE	Abbreviation for the American Society of Heating, Air-Conditioning and Refrigerating Engineers.
BTU	British thermal unit. The amount of heat required to raise the temperature of one pound of water one degree Fahrenheit at 4° Celsius. (39.2^{o}F.)

C	Celsius temperature scale wherein water freezes at 0° C and boils at 100° C at atmospheric pressure.
Calorie	The quantity of heat needed to raise the temperature of 1 gram of water 1° C.
Check valve	A one-way valve for he water going through a solar controlled hot water system (or pool). A check valve is used to prevent thermosiphoning.
Circumsolar Radiation	Radiation, coming from the Sun, that originates outside the solar disk.
Collector	A concentrator plus a receiver.
Collector Efficiency	The ratio of the energy collected by the solar collector to the radiant energy incident upon the collector.
Concentrator	Any device for gathering the Sun's rays and directing them in a way that produces useful, concentrated energy.
Conduction	The transfer of heat by contact with a hot body.
Dessicant	A chemical used to absorb contamination, including water vapor.
Diffuse Radiation	Scattered radiation from the Sun, as opposed to direct radiation, i.e. large angle scattered radiation with

the direct subtracted.

Direct Radiation — Radiation from the Sun measured at the Earth's surface that is received within a narrow solid angle.

Emissivity — The relative power of a surface to emit heat by radiation, or the ratio of the radiant energy emitted by a surface to that emitted by a black body--considered to have almost perfect heat absorption and therefore very low emissivity--when the given surface and the black body are at the same temperature.

Figure Error — In the fabrication of a flat or curved reflective surface, this term refers to the deviation of the real surface from its theoretical contour.

Flat Plate Collector — A collector that absorbs both the direct and solar beams. Usually, it consists of glazing, tubes, absorber plate and insulation.

Gasket — A piece of solid material, usually metal, rubber or plastic, placed between two pieces of pipe or between an automotive cylinder head and the cylinder block, for example, to make the metal-to-metal structure fluid tight.

Glazing — Glass, plastic or other transparent covering of a collector-absorber surface.

Gore	The name applied to a single section of a concentrator surface.
Heliostat	A flat reflector for directing the Sun's rays toward a fixed receiver.
Hole	A vacant electron state in the valence band of a material. It conducts electricity by acting like a positively charged electron.
Infrared Radiation	Electromagnetic radiation, whether from the Sun or a warm body, that has wavelengths longer than visible light.
Insolation	The total radiant flux of sunlight falling on a surface over a hemisphere. It is also referred to as total or global insolation.
Insulation	A material with high resistance or R-value that is used to retard heat flow.
Irradiance	The radiant energy falling on a unit area per unit time. It is usually stated in BTU/ft^2 or in $watts/m^2$.
Kilowatt	A measure of power equal to one thousand watts, approximately 1 1/3 horsepower, usually applied to electricity.
Kilowatt-hour	The amount of energy equivalent to

one kilowatt of power used for one hour--3,413 BTU.

Langley	The measure of solar radiation intensity equivalent to one calorie per square centimeter.
Lite	A section of glazing used in solar energy. Also called a Light or Blank.
Near UV	The wavelengths in the solar spectrum from approximately 200 to 400 nanometers.
Open Circuit Voltage	In a solar array, the voltage developed by the sunlight when the circuit is open. It is the maximum available in that circuit at a given irradiance.
Parabolic Concentrator	This refers to the mathematical shape of one type of concentrator. The equation of the surface is $y=r^2/4f$ where f is the focal length, r is the radius and y is the optical axis.
Passive System	A solar heating or cooling system that uses no external mechanical power to move the collected solar heat.
Peak Watt	A term used in solar electric power systems. When a system is rated at 1

peak watt, this means that it will deliver 1 watt at a specified working voltage under peak solar irradiance.

Phase Change — The change of state of a material, such as from a solid to a liquid.

Photons — Particles of light energy that transfer energy to electrons in photovoltiac cells.

Photovoltaic — The ability of a material to generate electrical energy when it is exposed to radiant energy such as sunshine.

Power Tower — A solar energy tower so placed that radiation from many heliostats surrounding it may be focused on the receiver at the top of the tower.

Pyranometer — A device for measuring global insolation. Also, it is referred to as global pyranometer or total pyranometer.

Pyroheliometer — A device for monitoring the solar intensity normal to the Sun's direction.

Pyrolysis — Decomposition of material, such as organics, into its basic chemical constituents by the action of heat.

Quad — The unit of energy to 10^{15} BTU. Sometimes defined as 10^{18} BTU.

Radiation	The flow of energy across open space via electromagnetic waves, such as visible light.
Radiometer	An instrument that measures any kind of radiation.
Resistance	The tendency of a material to retard the flow of heat, measured in (hr-ft^2 °F)/ BTU. Also known as R-value.
Selective Black Paint	More absorbent of the infrared long wavelengths of sunlight than non-selective black paint, hence, an improved material for coating the absorber plates in solar collectors.
Selective Surface	A surface that absorbs more energy than it emits.
Slope Error	The error in the angle or position of a surface from its expected position. It usually refers to changes of short range, such as equal to or less than one centimeter.
Solar Constant	The intensity of the Sun at the Earth. Typically, it is 1353 watts/m^2, 1940 cal/min-cm^2 or 429.2 BTU/ft^2-hr.
Solar Pond	A pond of stratified liquids designed to retain the energy from the Sun.

Special Distribution	A curve of the Sun's energy showing the variation of the irradiance with wavelength.
Spectral Reflectance	The ratio of the energy reflected from the plane surface for a given waveband to the energy incident.
Specular Reflectance	When solar energy is incident upon a surface, the reflected beam is broken up into two components: a narrow or specular component and a diffused or wide beam component. See diagram in the text.
Thermal Radiation	Electromagnetic radiation emitted by a warm body.
Ton of Cooling	12,000 BTU per hour. The term is derived from the amount of heat energy required to convert a ton of water into ice at 32 ^{o}F during a 24-hour period.
Ultraviolet Radiation	Ultraviolet radiation (UV) has wavelengths longer than those of X-rays but shorter than the visible. Usually this term refers to the wavelength region between 100 and 4000 angstroms One angstrom is 10^{-8} cm. The wavelengths of the UV radiation at the Earth's surface do not fall below 3950 angstroms (or 395 nanometers).
VDF	Vacuum-deposited film.

Watt — The energy rate of 1 joule (10^7 ergs) per second.

Chapter 8

PREFERRED SOLAR ENERGY REFERENCES

TEXTS

1. F. Kreith and J. Kreider, Principles of Solar Engineering, McGraw-Hill Book Co., Inc., 1978. Cost: $20.00.
2. D. Rapp, Solar Energy, Prentice-Hall, Inc., Englewood Cliffs, New Jersey 07632, 1981. Price: Approximately $18.00.
3. W.C. Turner, Editor, Energy Engineering Management Handbook, John Wiley & Sons, Inc., New York, NY 1982. Price: Approximately $49.50.
4. Bruce Anderson, (with Michael Riordan), The New Solar Home Book, Brick House Publishing Company, Andover, MA 1987.
5. Ted Lucas, How to Build a Solar Heater, Revised Edition, Crown Publishers, Inc., New York, NY 10016, 1975.

PERIODICALS

1. Journal of Solar Engineering, Produced by the ASTM, 345 East 47th Street, New York,

NY 10017. Price is approximately $60.00 per year for non-members.

2. Energy, The International Journal. Produced by Pergamon Press. The price is DM 1025 annually (1988).
   Available from:
   - Headington Hill, Oxford, OX3 OBW, U.K.
   - Fairview Park, Elmsford, New York NY 10523.
3. Solar and Wind Technology, An International Journal. Produced by Pergamon Press. Price: DM 330 annually (1989). Available from the same addresses listed in the previous addresses above.
4. Solar Age Magazine. It is currently out of production (since 1986). However, back issues are available for $3.50 each from SolarVision, Inc., Harrisville, NH 03450.
5. Journal of Spacecraft and Rockets (solar cells). Available from The American Institute of Aeronautics and Astronautics, 370 L'Enfant Ave., S.W., Washington, DC, 20024-2518. ($44.00/year to members, write for non-member subscription.

Chapter 9

SOURCES AND BIBLIOGRAPHY

In addition to the References, there are a vast number of books and literature that deals with the basic technology or applications of solar energy.

In this section, some of the more important background sources are presented.

Because of the proliferation of solar energy books, it is difficult to know which to choose. Much time can be wasted researching the various libraries. Most libraries only carry a few of the many books available. Therefore, this section has been included to point out literature that has been valuable to the author.

Concerning periodicals, the solar energy enthusiast or designer is not in such a favorable position. There are few solar energy periodicals available in the United States. Frequently, these are expensive and only available in a few universities. In addition, Many of the articles concerned with complex engineering systems designed for operation at a particular location on the Earth's surface. Often, an extensive background in mathematics and statistics is involved.

SOURCES AND BIBLIOGRAPHY

ASTRONOMY

1. Samuel Glasstone, Sourcebook on the Space Sciences, D. Van Nostrand Co., Inc., Princeton, NJ 1965.
2. Christopher Lampton, The Sun, Franklin Watts Publishing Co., New York, NY 1982.
3. Donat G. Wentzel, The Restless Sun, Smithsonian Institution Press, Washington, DC, 1989.

CLIMATE

1. Climatic Atlas of the United States, National Oceanic and Atmospheric Administration, Washington DC, 1968.
2. W. Sellers, Physical Climatology, University of Chicago Press, Chicago, IL, 1965.

CONCENTRATORS

1. Phil Heggen, Solar Concentrating Mirrors, Energy General, 158 Laurel Avenue, Menlo Park, CA 94025. ($40.00, 1990).
2. A. Rabi, Comparison of Solar Concentrators, Solar Energy, Volume 18, p. 93, 1976.

ECONOMIC ANALYSIS

1. Dr. Don Rapp, Solar Energy, Chapter 5 entitled Economic Analysis, Prentice-Hall, Englewood Cliffs NJ 07632, pp. 138-156, 1981.

ENGINEERING

1. Theodore Baumeister (ed.), Mark's Standard Handbook for Mechanical Engineers, Eighth Ed., McGraw-Hill Book Co., Inc., New York, NY, 1978.
2. Frank L. Bouquet, Introduction to Seals and Gaskets Engineering, Systems Co., 1988.

ENGINES

1. G. Walker, Stirling Cycle Machines, Clarendon Press, Oxford, England, 1973.

GLASS

1. Anon., Glass Factory, P.O. Box 1372, Tacos, NM 87571 (Strengthened Glass Catalogue).
2. Frank L. Bouquet, Solar Energy Simplified, Fourth Ed., Systems Co., P.O. Box 876, Graham, WA 98338.
3. Frank L. Bouquet, Strengthened Glass Technology, Systems Co., P. O. Box 876, Graham, WA 98338.
4. Alexus G. Pincus and Thomas R. Holmes, Annealing and Strengthening in the Glass Industry, 2nd Ed., 1987. (A scientific and engineering book, $22.00, 1991). Available: Ashlee Publishing Co., Inc., 310 Madison Ave., New York, NY 10017.

GREENHOUSES

1. Tim Magee, A Solar Greenhouse Guide for the Pacific Northwest, 2nd Ed., Ecotope Group, 2332 East Madison, Washington 98112. (92 pages) 1979.

HEATING AND COOLING

1. J.F. Freider and F. Kreith, Solar Heating and Cooling, McGraw-Hill Book Co., Inc., New York, NY, 1975.

2. ASHRAE Handbook of Fundamentals, American Society of Heating, Refrigerating and Air Conditioning Engineers, New York, NY, 1972.
3. Stu Campbell, Build Your Own Solar Water Heater, Garden Way Publishing Co., Charlotte, VT 05445 (109 pages) 1978.
4. James A. Autry, Energy-Saving Projects You Can Build, 1st Ed., Meredith Corp., DesMoines, Iowa (96 pages) 1978.
5. Sandra Oddo and Delores Wolfe, Sun Power: Keeping Warm the Solar Way, pp. 54-83, in Total Warmth, The Complete Guide to Winter Well-Being, Macmillian Publishing Co., Inc., 866 Third Avenue, New York, NY 10022 (210 pages) 1981.
6. Daniel K. Reif, Solar Retrofit, Adding Solar to Your Home, Brick Publishing Co., Andover, Mass. 01810, (197 pages) 1981.
7. Steven J. Strong and William G. Scheller, The Solar Electric House, A Design Manual for Home-Scale Photovolaic Systems, Rodale Press, Emmaus, Penn. (276 pages) 1987.

INSOLATION

1. Don Rapp and D. Oxley, On the Relationship Between Global Insolation on Horizontal and Tilted Surfaces, Energy Conversion, Volume 18, p. 39, 1978.

PASSIVE SYSTEMS

1. E. Mazari, The Passive Solar Energy Book, Rodale Press, Emmaus, Penn., 1979.

2. Darryl J. Strickler, Passive Solar Retrofit, How to Add Natural Heating and Cooling to Your Home, Van Nostrand Publishing Co., Reinhold Co., New York, NY (173 pages) 1982.
3. Brad Schepp and Stephen M. Hastie, The Complete Passive Solar Home Book, Tab Books, Inc., Blue Ridge Summit, Penn. 17214 (308 pages) 1985.
4. Helen Sweetland, Solar Remodeling, 1st Ed., Lane Publishing Co., Menlo Park, CA 94025 (96 pages) 1980.
5. Anne Coolman, Energy-Saving Projects for the Home, Orthobooks Co. (112 pages) 1980. Availability: Chevron Chemical Company, Consumer Products Div., San Francisco, CA, 91405.
6. Ted Lucas, How to Build a Solar Heater, Revised Ed., (242 pages), Crown Publishers, Inc., New York NY 10016.

RIGHTS, SOLAR

1. Dr. Donald Rapp, Solar Energy, Prentice-Hall, Inc. Englewood Cliffs, NJ 07532, 1981.

SELECTIVE SURFACES

1. A.B. Meinel and M.P. Meinel, Applied Solar Energy, Addison-Westley Inc., Reading, MA, ca. 1975.
2. B.O. Seraphin, Spectrally Selective Surfaces and Their Impact on Photochemical Solar Energy Conversion, Topics in applied Physics, Springer-Verlag, Berlin, Germany, 1979.

SUN

See Astronomy

TRANSMISSION THROUGH THE ATMOSPHERE

1. A.P. Thomas and M.P. Thekaekara, Experimental and Theoretical Studies on Solar Energy Conversion, Proc. 1976 Annual Meeting of the American Section of the International Solar Energy Society, Winnipeg, Canada, 1976.

RECENT BOOKS

BUILD YOUR OWN SOLAR HOT WATER SYSTEM ??

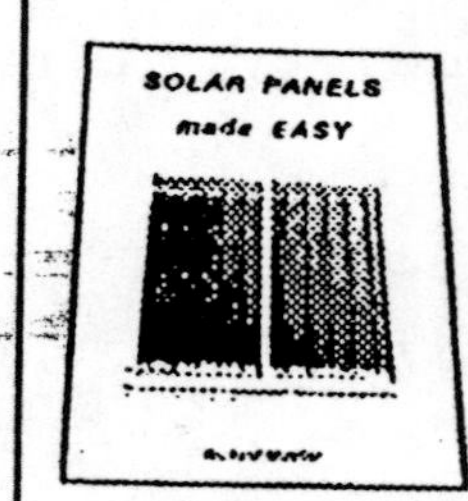

You bet! Learn how to build solar hot water panels by reading the book "SOLAR PANELS made EASY" from ***SOLAR SENSOR.*** Large 8½ x 11 inch format illustrated book describes how, in easy to understand language, with drawings, photos and dimmensions. Takes the guesswork out of designing and building your system. 30 day money back guarantee.

"SOLAR PANELS made EASY" book:
SP-1 $14.95 + $1.05 postage

LEARN THE BASICS OF PHOTOVOLTAICS

"The New Solar Electric Home" by Joel Davidson is a photovoltaics how-to hand-book written in non-technical terms, with lots of drawings and photos based on actual experience. Understand the benefits and limitations of photovoltaics by reading this 416 page book.

The New Solar Electric Home,
NS-1 $18.95 + $1.50 postage

PRACTICAL PHOTOVOLTAICS

by Richard J. Komp

This 200 page book is jam packed with useful information on how to test and use individual PV cells and build your own arrays. Included is information on how to calculate your load requirement, array size and battery bank. Read the technical section that describes how PV cells work, how to solder them together and how they are manufactured. This book is **"A MUST"** for experimenters working with PV cells.

PP-1 $16.95 + $1.05 postage

RV'ers GUIDE TO SOLAR BATTERY CHARGING

Answers many questions about RV battery charging, use of appliances with a PV system. Suggests good places to install batteries, panels and other equipment. Defines all the electrical terms used in PV system design and basic electricity. Covers RV space heating and cooling techniques, use of refrigerators, TV's, computers and other appliances. Describes actual systems being used today. Includes wiring diagrams.

RV-1 $12.95 + $1.05 postage

SOLAR SENSOR
10909 Hayvenhurst Avenue
Granada Hills, CA 91344

INDEX

ABOUT THE AUTHOR

Frank L. Bouquet was born in Oregon and grew up in the Midwest. He was active in many phases of Scouting and then served in World War II and the Korean Police Action.

His B.A. degree was from the University of California (1950) and M.A. degree in Physics was from U.C.L.A. (1953).

As a physicist for the U.S.Navy, he analyzed the fallout spectra from the first U.S. hydrogen bomb (Mike). Later he held various positions including Acting Manager of Lockheed Nuclear and Space Group. At NASA (JPL), he was involved in the materials and radiation design of spacecraft.

He held various positions in the IEEE Nuclear and Space Society (Los Angeles Chapter) and was a member of the AIAA. He has published 73 technical papers and over 100 government and industry reports. Listings include Who's Who in Aerospace, America and the World. He is an active consultant living in Oregon.